INTELLIGENT LEARNING SYSTEM

SUPPORTING LEARNERS

Muhammad Adnan

Dedication

To my parents, whos' prayers are always with me and had sacrificed a lot to see me succeed in life.

To my wife, who provided me every assistance during the time I need them and who think that my study will never complete.

To my brothers, Roshan, Rehan, and Naveed who helped me in selecting the right direction and supported me in every possible way.

To my sisters, Dr. Abida, Dr. Sajida, and Dr. Saima who proofread my work and helped me in carrying out experiments and generating results.

To my lovely sons and daughter who made me happy at the time of agony

Abstract

Machine learning and deep learning algorithms work well where future estimations and predictions are required. Notably, in educational institutions, machine learning, and deep learning algorithms can help instructors predict learners' learning performance. Furthermore, the prediction of learners' learning performance can assist instructors and intelligent learning systems in taking preemptive measures/early engagement/early intervention measures so that the learning performance of weak learners could be increased, thus reducing learners' failures/dropout rates.

In this study, we propose an intelligent learning system that provides adaptive feedback and support to learners. The main component of an intelligent learning system is the mobile -learning (m-learning) model. Different m-learning models are created for diverse learners according to their learning features (study behavior) and feature weights. Before creating the m-learning model, learning features are elicited from learners' mobile learning activities and stored on persistent online cloud storage. For a broader understanding of learners' learning behavior, learning features were divided into three categories: context features, behavioral features, and final performance features. Context features manifest how the learner will use mobile devices as a learning tool in different contexts. Context features include knowledge background, learning time, learning time duration, places visited, learning content type (video, text), learning performance, etc. Behavioral features illustrate how the learner interacts with the mobile device during the learning process. Examples of behavioral features include online learner participation in discussion groups, the number of times problems posted online, the number of times problem solved, topic repetition rate, and the number of times attempted the quiz. Lastly, the final performance feature explains the learners' performance in the final learning activity. The final performance feature is the dependent feature, and context/behavioral features are the independent features. In this study, we are interested in revealing how each independent feature contributes to increasing a learner's final performance. Knowing independent feature importance/weights in improving final performance allows intelligent learning systems/instructors to provide timely adaptive help, support, and feedback. Moreover, an intelligent learning system could take early engagement/early intervention measures to reduce dropout or failure rates.

After features analysis and preprocessing steps, we created an m-learning model to categorize learners into performance groups (grades). Because of the high accuracy of deep learning algorithms, we used Artificial Neural Network (ANN) to classify learners into five performance groups, whereas Random Forest (RF) algorithm was used to determine each feature's importance and weight in the creation of the m-learning model. ANN-based m-learning model is dynamic, which means that once fully trained, the model can predict new learner performance grades given his/her learning features. To sieve the m-learning model from overfitting and underfitting problems, we trained it on learners' features instances iteratively until we gained maximum accuracy for learners' grades prediction. For the training ANN-based m-learning model, we used techniques such as forward propagation, back-propagation, ReLU, and Softmax activation function. We used the stochastic gradient descent technique for minimum error/cost in the model. It is more efficient and impressive when dealing with high dimensional data, and ANN synapses need suitable weight values for generating a minimum error.

In the last stage of this study, we performed an early intervention/engagement experiment on those learners who showed weak performance in their study. The early intervention experiment's objective was to encourage and help vulnerable learners increase their learning behavior by sending motivational and adaptive triggers/messages on their mobile devices. For the early intervention experiment, 425 learners were selected (learners having D and F grades). They were further divided into the control group and experimental group learners (the control group had 212 learners, whereas the experimental group received 213 learners). Two types of triggers were sent on experimental group learners, namely adaptive and motivational triggers. In contrast, the control group learners were independent of receiving any adaptive content and motivational/adaptive triggers. The early intervention experiment lasted for 50 days. Its result concluded that intervened learners showed overall better performance than un-intervened learners in the final learning activity (Intervened learners had overall 12.30 percent higher performance than un-intervened learners). Lastly, we used the End-User Computing Satisfaction (EUCS) model questionnaire to measure learners' attitude towards using an intelligent learning system. EUCS questionnaire was shared only with 213 experimental group learners as the control group was independent of using an intelligent learning system during the early intervention experiment. Six dimensions of client-side application, namely learnit, were used to determine learners' satisfaction level towards it. The six dimensions of EUCS included usefulness, ease-of-use, engagement, timeliness, adaptiveness, and attitude towards learning software applications. Overall, the scores recorded for all six dimensions of EUCS were greater than 4, which inferred that overall, the experimental group learners were satisfied with intelligent learning system functionalities and will use such software applications in the future.

Contents

List of Tables

List of Figures

Acronyms

ILS Intelligent Learning System

ITS Intelligent Tutoring System

ITT Intelligent Tutoring Tools

DM Domain Model

LM Learner/Learning model

TM Tutoring Model

UI User Interface

UIM User Interface Model

E-Learning Electronic Learning

LMS Learning Management System

MOOC Massive Open Online Course

AL Adaptive Learning

MF Mobile Learning

ML Machine Learning

DL Deep Learning

KNN K-Nearest Neighbors

TAM Technology Acceptance Model

HL Hidden Layers

OL Output Layers

CAL Computer-Aided Learning

CCL Cross-Contextual Learning

PaaS Platform as a Service

SaaS Software as a Service

BKT Bayesian Knowledge Tracing

MT Model Trees

LDA Linear Discriminant Analysis

MCLR Multi-class Logistic Regression

SVM Support Vector Machine

DT Decision Trees

RF Random Forest

ANN Artificial Neural Network

AL Artificial Learning

AI Artificial Intelligence

CNN Convolutional Neural Network

RNN Recurrent Neural Network

SOM Self-organizing maps

DBM Deep Boltzmann Machine

DAE Deep auto-encoder

PC Personal Computer

MD Mobile Device

EUCS End-User Computing Satisfaction

DNN Deep Neural Network

IL Input Layers

CBT Computer-Based Training

CBL Computer-Based Learning

DL Distance Learning

SL Self Learning

PDA Personal Digital Assistant

BaaS Backend as a Service

NFC Near Field Communication

IRT Item Response Theory

LWLR Locally Weighted Linear Regression

DBSCAN Density-Based Spatial Clustering of Applications with Noise

GPU Graphics Processing Units

LSTM Long Short Term Memory

SGD Stochastic Gradient Descent

BPNN Back-Propagation Neural Networks

HMM Hidden Markov Model

RMSE Root Mean Square Error

SDK Software Development Kit

ODGP Online Discussion Group Participation

NTPS Number of Times Problem Solved

TRR Topic Repetition Rate

NTPP Number of times problem posted

NTRA Number of Times Text Resources Accessed

NTAQ Number of Times Attempted Quiz

AST Average Study Time

NTVRA Number of Times Video Resources Accessed

APV Academic Places Visited

FG Final Grades

MP1 Learner Performance in Module 1

MP2 Learner Performance in Module 2

MP3 Learner Performance in Module 3

SPV Social Places Visited

JSON JavaScript Object Notation

EM Expectation-Maximization

RLU Rectified Linear Unit

MSE Mean Squared Error

BLSTM Bidirectional LSTM

GNB Gaussian Naïve Bayes

MAE Mean Absolute Error

BPTT Backpropagation Through Time

CSV Comma Separated Values

FBM Foggs Behavior Model

INTRODUCTION

1.1 Intelligent Learning System and Mobile Learning

An intelligent learning system (ILS) considers learners' behavior before providing feedback, assistance, and learning content. ILSs adapt to the learning flow or pace of the learners. The learners' learning behavior is made up of learning features generated during the learning process. Learning features include study time, location, background knowledge, performance, social interaction, preferences, etc., of learners, [1]. ILSs must know the real values of learning features as learning content and feedback are given to learners by ILSs according to these features' real values. Unlike traditional classroom settings, ILSs provide customized instructions to different learners without intervention from human instructors. ILS uses a variety of computer technologies to make learning more effective and meaningful. The application of ILSs is everywhere. They are used in formal learning environments, informal learning environments, professional training, flight simulations, technical learning environments, etc. [2]. The implementation of ILSs has demonstrated its strength and weaknesses [3]. The ongoing research in academic settings and industries tries to improve the effectiveness of ILSs. Unlike traditional classroom settings based on a one-size-fits-all approach, ILSs try to provide one-to-one, personalized, and adaptive tutoring, thus making learning more engaging and exciting. ILS, also known as Intelligent Tutoring System (ITS), Intelligent Tutoring Tools (ITT), comprises four basics components based on general agreement amongst researchers.

- The domain model
- The learner model
- The tutoring model
- The user interface (UI) model

The learning objectives of the subject of interest, along with its learning content, are defined in the domain model [4]. The domain model also considers the necessary steps that are required for understanding a problem. In ILSs, the domain model is adaptive, designed to reflect the knowledge level of the learners. More specifically, a comprehensive domain in ILS defines the learning objective, the knowledge required to achieve that learning objective, learning content comprehension, analysis, application, and evaluation. In an adaptive domain model, learning content is designed according to the knowledge level of different learners; thus, in one domain model, different learning content is designed reflecting the knowledge level of diverse learners [5].

The core component of ILS is the learner model. ILS pays special attention to learners' performance state, background knowledge, and how the learners' knowledge evolves in the learning process [6]. Step-by-step guidance is provided to learners by ILS based on the learner model. Whenever the learner deviates from the domain model, the ILS notifies the learner that s/he has shifted from the learning objective. As the learning process progresses, the learner model stores information about the learner's understanding of its content. A more comprehensive learner model is formed when the learner frequently interacts with ILS. ILS then uses this learner model to deliver appropriate instructions and content to the individual learner [7].

In ILS, the tutoring model is the middleman taking input both from the domain model and the learner model and, based on these inputs, make choices about learning strategies and actions [8, 9]. The tutoring model tries to provide timely guidance and feedback to learners to achieve the desired learning objective in less time. At any time during the learning process, the learner may request guidance or what to do next, according to their proficiency and skills. The outcome of the tutoring model is to decide whether the learner is in a learned state or unlearned state. During the learning process, the tutoring model keeps track of learner's progress from problem to problem and develops learners' strengths and weaknesses.

The user interface (UI) model connects the learner model with the domain model via a software application, i.e., web application, desktop application, mobile application, etc. [10]. The learning objectives and learning content material are delivered to the learner via the UI model. Moreover, the UI model provides customize software interfaces to learners according to their interaction with the ILS. The UI model's goal is to provide an interface to the learner that is easy to use and enables learners to accomplish her/his desired learning task.

For the last three decades, ILSs have been implemented in a variety of forms. In classroom settings, ILS is implemented as desktop applications to support traditional learning where a learner is guided and supported by computer-assisted instructions. When implemented in the form of electronic learning (e-learning), distance learning, etc., ILS does not require instructors and learners to be present at the same time and location. Instead, instructors and learners communicate via wired and wireless infrastructure. In comparison to desktop application systems, the popularity of e-learning has increased rapidly to support learners using learning tools such as Learning Management Systems (LMS), e-mails, learning content, and websites [11]. Recently, Massive Open Online Courses (MOOCs) have gained tremendous popularity, and success [12]. MOOCs are based on an e-learning platform where learning content and education material are widely available to everyone. Whether it is LMS or MOOCs, the learner only needs to login into the system to access learning content. Most LMS and MOOCs follow a one-size-fits-all approach where learning content and guidelines are not adaptive [13]. The same learning material is available to all learners regardless of their learning behavior and performance state. The challenge for the current educational learning environments is to make learning smart and adaptive where learning content and instructions adapt to the learning behavior of the individual learner [14].

Adaptive learning (AL) is an educational method that uses computer algorithms to orchestrate the interaction with the learner and delivers tailored learning content according to the needs and performance of each learner [15]. In AL systems, learning content is presented according to each learner's unique needs as indicated by learner responses, interactions, and feedback. The need for AL has been driven by the realization that adaptive learning content cannot be delivered on a large-scale using the traditional and non-adaptive mechanisms. When properly implemented AL system can transform the learning process from passive to active learning. Though the primary target of AL systems is education, they can be used in various areas, i.e., professional training, business education, simulation, etc. AL systems have been designed primarily for desktop applications, web applications and recently have been the prime target for mobile learning systems.

The use of mobile phones for learning purposes is called Mobile learning (M-learning) which now is the prime target of ILS [16]. The use of mobile phones for e-learning, u-learning, M learning creates additional challenges for the delivery of adaptive learning content [17]. First, the limited space on a mobile phone screen does not allow learning content to be presented in the same way as shown on web and desktop applications. Moreover, due to low and inconsistent Internet speed, mobile learners feel frustration when accessing learning content during the learning process. Nonetheless, M-learning has several advantages over traditional and web-based learning systems. M-learning-based education has already reached many learners as it provides the possibility of an adaptation process that is difficult to achieve in conventional classroom settings [18]. Mobile learners (M-learners) using mobile phone technologies can access and study learning content at any time and from anywhere. Thus, M-learners are not limited only to classroom learning. They have the freedom to learn unrestricted time and location constraints. In m-learning settings, m-learners have the freedom to choose the learning content they want to know, unlike traditional classroom settings where learners are bound to the instructor's provided learning content and instructions. 'On the move' learning allows the M-learners to immediately apply the accessed information in their learning settings rather than waiting to use the data later at some time. Despite the numerous advantages of M-learning, M-learning's practicality had been limited due to not knowing about the exact behavior or learning features of M-learners and how these learning features have different meanings and values for different M-learners. Eliciting proper and right learning features in M-learning becomes more challenging as learning occurs across multiple contexts with diverse learning content and social interaction [19].

ILS commonly uses M-learning for informal learning, self-learning, or u-learning. In ILS, creating a learning model is a crucial step as the adaptation process is dependent on it [20]. The learning model is made up of learning features that are produced during the learning process. Learning features include learning time, learning context, learning time duration, preferred learning content (video, text, quizzes, etc.), learner performance, preferences, and content repetition rate, etc. [21]. Devices such as computers and mobile phones are initially not programmed to interpret learning features and context information the same way humans do during the learning and interaction process. For transparently using learning features, these devices must store, analyze, interpret, encode, and standardized learning features to respond and communicate appropriately. In M-learning applications learning context and personal learning features have a strong relationship. Personal learning features define the learning behavior, and learning behavior determines the learning context of the learner. The adaptation concept comprises two elements: personalization related to M-learner and automatic learning content customization about the adaptation process.

Personalization related to M-learner means that ILS knows the learning features, their importance, features weights of each individual and adjusts to M-learner's learning flow accordingly [22]. Adaptation means that ILS creates a learning model based on the learner's learning features automatically (machine learning) to provide the support that suits the M-learner's needs and situation. The adaptive learning system frequently extracts, interprets, and uses learning features to adapt its services to learners' current learning situations. While the concept is simple, such an adaptive learning system's development becomes challenging due to a lack of proper learning features dataset and their real values [23]. Moreover, the development of an adaptive M-learning system becomes more challenging as the learner is on the move, and the context changes frequently. In addition to extracting and processing learning features, the m-learning system must reason and decide how to process learning features and deliver personalized, meaningful, and useful services to m-learners. In an adaptive learning system, the most challenging issue is adaptation due to imprecise learning features and rapid changes in learning features. Due to this reason, some vital information in the learning features dataset generated by the learning system is left unexplored. As a result, the exact learning behavior or needs of learners is often overlooked. The challenge for effective ILS is to provide the right and suitable learning content at the right time and the right way.

1.2 Machine Learning in Intelligent Learning Systems

ILSs seek to draw conventions about individual learners based on their learning features and interaction with the system. ILS integrated with machine learning (ML) techniques can be used to create a learner model that properly reflects the learning behavior of individual learners [24]. A carefully developed learner model based on proper learning features can predict and identify learner strengths, weaknesses, preferences, and performance. ML techniques that have been used to developed learner model include K-Nearest Neighbors (KNN) [25], Linear Discriminant Analysis (LDA) [26], Multi-class Logistic Regression (softmax regression) [27], Support Vector Machine (SVM) [28], Decision Trees (DT) [29], Random Forest (RF) [30], and Artificial Neural Networks (ANNs) [31]. Fundamental steps in developing a learner model include defining proper learning features (time, location, preferences, performance, etc.), generating a dataset of learning features, features analysis, preprocessing, transformation into the appropriate form, choosing proper machine learning algorithm, training, testing and re-evaluating the learner model. When developed and matured, the learner model can be used as a part of ILS for intelligent guidance and adaptive support. The learner model can tell us the current learning behavior and the predicted learning behavior of individual learners. Based on this fact, the likelihood of achieving learning outcomes can be determined. The learner model can recommend additional learning content during the learning process if they need it and let the learner move ahead if the current topic is already covered. An adaptive M-learning system performs adaptation based on the learner model and updates the learner model when new facts are driven from M-learners' learning features. Information gathered from M-learning systems' interaction mechanism is often imprecise, and its interpretation is typically uncertain. Machine learning algorithms are commonly used for learner modeling because of their ability to represent complex feature relationships, feature weights, and combined features' effects on individual learners' learning behavior in the learning environment.

Machine learning (ML) is a subset of artificial learning (AI) which gives the computer the ability to learn without being explicitly programmed [32]. ML algorithms build a mathematical model based on training data to make decisions or predictions without being explicitly instructed to perform that task. With ML algorithms' help, ILSs can be designed to automatically extract meaningful patterns from learning features data in educational settings. Whether the data generated from learning features are fine-grained or less fine-grained, ML techniques leverage both types of data to discover meaningful patterns about different kinds of learners and how they study, structure their course, and the effect of instructional strategies embedded in the learning environment. These learning patterns provide new insight into learners' learning behavior that would be very difficult to discern by looking at the raw data. As the research and development in ILSs continued to grow, many ML techniques could be in a variety of educational contexts. ML techniques can help ILSs translate raw data into meaningful information related to the learning process to make improved decisions about the trajectory and design of learning objectives. For ML algorithms, the provision of proper learning features and their corresponding instances is the key step [33]. Once suitable learning features are provided, ML algorithms can discover the relationship in learning features. This involves searching through a repository where learning features are stored to find consistent relationships among learning features. Based on these learning feature relationships, ML algorithms can perform feature classifications, clustering, regression, social network analysis, and association rule mining, etc. [34]. After finding feature relationships, ML algorithms can make predictions about future events in the learning environment. Predictions can help ILSs in making policy decisions and decision-making processes.

Deep learning (DL) is a sub-field of machine learning that gives the computer the ability to adapt to new data without human intervention [35]. Like machine learning algorithms, deep learning algorithms extract new information from current features for decision-making and predictions. DL algorithms are used to learn feature hierarchies, weights, and interrelationships often based on Artificial Neural Networks (ANNs). Models based on DL algorithms can focus on the right features by themselves, requiring little intervention from human programmers [36]. ML and DL algorithms have the single most important AI research goal: allowing computers to understand and model our real world well enough to exhibit something identical to what we human beings call intelligence. Some of the popular DL algorithms are Artificial Neural Network (ANN) [37], Convolutional Neural Network (CNN) [38], Recurrent Neural Network (RNN) [39], self-organizing maps (SOM) [40], deep Boltzmann machine (DBM) [41], and deep autoencoder [42].

The use of the DL technique for personalized M-learning is an emerging field aimed at optimizing the pace of learning and instructional sequence according to the needs of each learner [43]. The learning features (learner performance, preferences, content types, learning time, learning time duration, learning location, repetition rate, social interaction, etc.) based dataset act as a fuel to DL algorithms based on which DL algorithms can classify M-learners into different learning groups having differential interests and needs. Sequencing instructions and course contents are essential in the m-learning process as M-learners get a lot of random information from the Internet, Learning Management System (LMS), and mobile devices communications. Nevertheless, the development of the DL-based M-learning model is non-trivial for every m-learner and can help them take the right learning path. Now m-learning has become a common learning trend for the last decade, and there has been a rapid increase in the numbers of m-learners in colleges and universities who access information from the Internet to get learning help. Yet, on the other hand, m-learning is still in infancy. Therefore, an effort is needed to investigate how to change the learning paradigm from a one-size-fits-all approach to adaptive and personalized learning supported by mobile devices. M-learning is inevitable as two-thirds of the learners are already on smartphones making it a handy medium of gaining knowledge [44]. Mobile devices provide a more flexibility level, and a course started on a PC or tablet can be completed on a mobile device. Moreover, learning is more contextual and heuristic because learners are on the move and learning can take without spatial and temporal constraints [45]. In m-learning settings, creating a learner model and modeling learners accordingly to it is the most potent apparatus for successful and effective adaptive m-learning.

1.3 Problem Statement and Objectives

In the m-learning environment, a research challenge exists to develop a learner model that can deliver tailored learning content to the individual learners considering their learning behavior features. Developing a learner model that intelligently selects learning resources for individual learners to enhance their learning capacity is the prime need for adaptive learning environments. Moreover, adaptive learning environments also need learner models that recommend learning content to learners with high efficiency and accuracy. To enhance the accuracy and efficiency of the learner model, different machine learning (ML) and deep learning (DL) algorithms exist that can help the learner model make the right decisions with less time. It is also challenging to analyze and select the best ML and DL algorithms for adaptive learning systems. In adaptive learning environments, the learner model must control learners' learning progress according to their learning capabilities. Implementing the right ML and DL algorithms can make a drastic transformation in how a learner can develop and learn. Considering the challenges and problems, below are the objectives of this research thesis.

- To develop an intelligent learning system (ILS) for M-learning based on significant learning features that contribute most to the M-learning process.

- To develop a learner model (M-learner model, adaptive model, learning model) based on significant learning features as a part of larger ILS.
- Using ML and DL workflows to process and transform learning features to create a learner model.
- To develop an M-learner model that will predict learners' performances with maximum accuracy and precision using the weight difference among learning features.
- To test and verify the learner model-based ILS's effectiveness by state-of-the-art evaluation method, i.e., End-User Computing Satisfaction (EUCS).

In the m-learning environment, it is often difficult to determine which features make up a learner model and where these features should be logged or stored. Moreover, it is also challenging to develop a learning system that perfectly represents learners' interaction with the learning system.

1.4 Research Contribution

This study's major research contribution is the development of an Artificial Neural Network (ANN) based intelligent learning system that provides adaptive support and guidance to learners according to their learning behavior. The main component of an intelligent learning system is the m-learning model that ANN creates through learning features. The m-learning model is created by considering two factors: 1. Important learning features that define the learning behavior of m-learners. 2. What is each feature's weight in determining the learning behavior of m-learner? An important disadvantage of contemporary learning models is that they do not consider weight differences among different learning features while defining the learning behavior for different m-learners. Therefore, in this study, an efficient and powerful m-learning model is proposed and presented that intuitively classify m-learners according to their learning features, weights, and interrelationship.

In this study, the m-learning model is introduced and developed particularly for the m-learning system, online learning, and intelligent learning systems that use mobile devices for user interaction. The working of the m-learning model is based on a deep learning algorithm (deep neural network). For the adaptive m-learning model, the most important component is the learner dynamic learning context. One of the most significant applications of our m-learning model is to create a knowledge model of learners that precisely represents their performance factors and learning context. The importance of learning features, relationships, and weights are different for different learners in diverse contexts. Precisely the contributions of this study are:

- Creating an m-learning dataset. The dataset includes learning features instances/records generated during the learning process and logged on an online Firebase cloud. The dataset is used as an input to ANN for the development of the m-learning model.

- M-learners behavior modeling. We used the m-learners' dataset to model learning in the m-learning environment i.e., creating the m-learning model. In modeling m-learners, the important step is to determine the effect of independent learning features on the dependent feature, i.e., m-learners performance (final grades).

- Using Artificial Neural Network (ANN) for learner's classification. ANN (with hidden layers) is a type of deep neural network or deep learning algorithm used for m-learners classification, important features identification (learning features), and feature weight determination.

- Deployment of the m-learning model on mobile devices. In this step, we deployed the m-learning model on mobile devices of m-learners to determine how accurately the learning performance of m-learners is predicted and how the m-learning model helps m-learners in the learning process and making the right decision.

- Performing early intervention/engagement experiment. This experiment's objective was to determine whether early intervention by an intelligent learning system motivates m-learners to improve their performance. This experiment revealed that early intervention could be a beneficial technique in encouraging weak learners to enhance their learning behavior.

Moreover, the key considerations of the proposed work are listed in table 1.1.

Table 1.1: Key considerations of the proposed work

Key Considerations	Current Literature	The Proposed Work
Consideration of learners' learning features	Most of the articles related to machine learning do not target attributes weights and their interrelationships	Important features like learners' performance, preferences, time, learning time duration, and environmental context will be considered in the adaptive learning mechanism
Mobile learning dataset	There is no proper learning dataset that records mobile learning behavior during the learning process	Data related to mobile learning behavior will be logged on the cloud for input to the learner model
Analyzing applications and performance of deep learning models	Classical and traditional machine learning model is used in adaptive online (web-based learning) and M-learning	Analysis and implementation of cutting edge, state-of-the-art deep learning algorithms in predicting learner knowledge model
Deep neural network support on mobile devices	Mobile devices currently are used for image recognition, character reading, etc.	Deployment and testing of the learner model on mobile devices

1.5 Organization of Thesis

The rest of the thesis is organized in the following chapters:

Chapter 2: In this chapter, the current literature about applying machine learning and deep learning techniques in educational settings are discussed, emphasizing adaptive learning systems. This chapter starts with a broader explanation of learning technologies, distance learning (d-learning), electronic learning (e-learning), mobile learning (m-learning), context-aware learning, and it narrows towards more related work done in the application of machine learning and deep learning in the adaptive learning environment. This chapter also covers a detailed description of machine learning and deep learning algorithms and how they can be trained and used in intelligent learning systems. Notably, we highlighted the most popular and common studies that discuss the application of machine learning algorithms such as Decision Trees (DT), Support Vector Machines (SVM), Locally Weighted Linear Regression (LWLR), K-Nearest Neighbors (K-NN), and Random Forest (RF), etc., in the learning domain. At the end of this chapter, we discussed various deep learning algorithms that are based on the neural network architecture (Artificial Neural Networks (ANN), Convolutional Neural Networks (CNN), Recurrent Neural Networks (RNN), etc.,) and their related studies that explain how they can be applied in learning domains.

Chapter 3: In this chapter, we discussed the complete working of the proposed intelligent learning system. Firstly, we discussed how learning features were collected from m-learners and were stored on an online database cloud called Firebase for Artificial Neural Network (ANN) processing. For learning features collection, we used an Android app called learnit that interacts with the learners and presented different learning content related to computer programming (JAVA, C++) and computer basics. While m-learners interact and use the learnit app, their learning features are recorded, logged, and stored on the persistent Firebase online database. Later, these learning features are preprocessed and feed into ANN for the m-learners performance classification task.

Chapter 4: This chapter explains how ANN is trained to make the m-learning model and is deployed on mobile devices to guide m-learners in their learning process. This chapter further elaborates on how feature weights/importance is determined using the Random Forest (RF) ensemble technique. The feature weight values are used for providing adaptive and tailored support to different m-learners. Later, we discussed how early intervention could motivate m-learners in increasing their learning behavior. At the end of this chapter, we discussed the End-User Computing Satisfaction (EUCS) model and how we used it in determining the attitude/satisfaction of m-learner in using intelligent learning systems.

Chapter 5: This chapter concludes our research study and gives future direction on applying machine learning and deep learning techniques in educational settings.

LITERATURE REVIEW

In this chapter, the fundamental concepts associated with online learning and its technologies are discussed. This chapter's flow starts from the broader concepts related to online learning, e-learning, mobile learning (M-learning) and narrows down to how machine learning techniques could be used to increase their usefulness. This chapter highlights the transition from e-learning to m-learning and describes the concepts related to learner modeling, adaptation, personalization. This chapter ends with descriptions of techniques used in adaptive learning and, more specifically, review the literature that discusses machine and deep learning techniques used in adaptive learning.

2.1 Learning Technologies

The main goal of learning technologies is to assist both instructors and learners in making the learning process easy, engaging, interesting, and productive. However, the use of learning technologies in educational settings was minimal until recently [46]. In the past decade, the application of learning technologies has successfully enhanced assessment, learning, and teaching experience. Learning technologies include computers, tablets, laptops, mobile devices, computer networks, communication technologies, and multimedia technology to support learning. In the past, learning technologies have been classified under various acronyms such as computer-based training (CBT) [47], computer-aided learning (CAL) [48], computer-aided instruction (CAI) [49], computer-based learning (CBL) [50], and computer-aided assessment (CAA) [51], etc. With the advancement in learning technologies, the growth of distance learning (D-learning) and electronic learning (e-learning) continue to increase, aiming at supporting learning participants through the use of learning tools such as virtual classrooms, emails, websites, electronic learning content, etc. Learning technologies also help learners in making the learning experience more personalized. Learning technologies normally make the learning experience personalized by recommending appropriate learning content to the learner. With the emergence of machine learning, learning technologies could be used to create personalized learning content and learning flow. Learning technologies mainly rely on data for adaptive learning content recommendation. Enough data collected from learners during the learning process and sufficient data collected from the nature of learning content can enable learning technologies to recommend adaptive learning experiences. Learning technologies tools such as Flipboard, Google classroom, schoololgy.com, cloud-based word processors, etc., have removed classroom boundaries where learners and instructors can collaborate in real-time unrestricted time, and spatial constraints [52]. The advancement in learning technologies has helped in mobile learning's rapid growth (M-learning) and distance learning (D-learning). The use of the Internet has enabled instructors to reach learners across the borders, and learners from any country can register for online virtual classrooms. Learning technologies have made education easy, affordable, and flexible where learners can take classrooms during their free time free of cost and learners even can interact with other learners virtually. Learning technologies have digitized traditional classrooms with learning tools like smart electronic whiteboards, learning management systems (LMS), computers, laptops, pads, tablets, smartphones, enabling increased learners' engagement and motivation towards learning. Learning technological tools such as mobile devices and the Internet gives the learner the chance to learn by him/herself and gives the learner the freedom to discover learning content according to her/his needs and preferences.

2.2 Distance Learning (D-learning)

In D-learning, a learner can take college/university courses from anywhere [53]. A learner can study from home, library, or anywhere without needing to come to college or university. There is no identification and age restriction in enrolling in D-learning courses. The advancement in communication and transportation technologies has made possible the implementation of D-learning. Communication technologies are used to bridge the gap between instructor and learner as face-to-face interaction is rare in D-learning. Apart from a much higher degree of freedom, D-learning is very much like traditional education as it covers all degree levels and subjects offered in colleges and universities. A learner is D-learning taught by an instructor, receives support from a tutor, takes examinations, and discusses the problems with classmates without a tight schedule and physical presence constraints. D-learning gives the learners the advantage of learning at their speed, from anywhere and at any time. Massive open online courses (MOOCs), now offered by many universities, provide open access to learning content through the World Wide Web (WWW) are recent developments in D-learning

[54]. Generally, terms such as online learning, virtual classroom, distributed learning, e-learning, etc., are used interchangeably and synonymously with D-learning. The widespread use of the Internet has made D-learning easy and affordable. Today, multitudes of virtual schools, virtual colleges, and virtual universities offered full degree programs covering almost all subjects. Earlier D-learning was supported by technologies such as telephones, video conferencing, television, the radio, and text-based education, which have become somewhat redundant with the invention of the Internet. Advancement on the Internet and wireless technologies have made D-learning more attractive and easier with the support of video, voice, text, multimedia content, and immersion teaching methods.

2.3 Electronic Learning (e-learning)

In D-learning, the distance between learners and instructors is a factor, whereas, in e-learning, distance is not much of the factor [55]. E-learning is a supportive tool for both instructors and learners, which helps both inside and outside the classroom. E-learning is a kind of learning technology that allows the learner during his/her learning process. Most people do not try to realize that e-learning can also happen right inside the class. E-learning combines digitally delivered learning content with learning support and services. All types of electronic media, including the Internet, extranet, intranet, cloud support, audio/video files, and satellite broadcast, are used in e-learning to deliver learning materials to learners. The term e-learning and online learning are almost the same having the following characteristics:

- E-learning enables learners to take control and access what they need quickly and easily.
- When it happens outside of a classroom, E-learning encourages the learner to do independent learning, hence not waiting for the instructor's instructions to move ahead.
- The learning content provided for any targeted topic can be seen and rewind again and again until the topic is wholly understood.
- E-learning enables instructors to ensure that learners are always synchronized

with the latest learning material. This fact increases the learner's motivation by realizing that s/he is still accessing updated content.

- E-learning promotes easy and quick learning. E-learning helps instructors create and communicate new lessons, contents, ideas, policies, and instructions to learners, whether formal or informal.
- E-learning enables instructors to deliver the same learning material in a consistent way to all the target learners. This ensures that all learners are treated equally and receives the same type of learning material.
- As compared to traditional learning, e-learning is cost-effective as learning in this mode happens quickly, easily, and effortlessly.
- As e-learning is supported by audio, video, and multimedia aid, it is easy to grasp and understand the learning content, which positively influences educational institution profitability.

Research studies in [56, 57], proved that e-learning as compared to traditional learning has the following advantages:

- It results in improved scores on quizzes, certifications, and other types of evaluations.
- Increased learners' engagement in online courses and a higher number of learners achieve a pass or good score.
- E-learning tools and technologies help the educational institution retain learning content for a more extended period.
- E-learning tools give learners an enhanced ability to learn and implement new knowledge at any time and from anywhere.

2.4 Mobile Learning (M-learning)

With the emergence of mobile devices (tablets, mobile phones, smartphones, etc.), learners have shifted their learning experience to a broader level. Mobile devices have enabled learners to learn in diverse contexts, including continuous change in their learning location and environment. ShengQuan Yu et al. explains that M-learning is "not primarily for delivering learning content, courses and messages but to involve learners in the broader learning experience and involvement through social and content interaction across multiple learning contexts" [58]. Occasionally mobile devices are used for informal learning, but it's a challenge to attract learners to do formal learning using mobile devices [59]. M-learning gives the learner the freedom to access learning content from virtually anywhere. Sharing ideas and problems is almost instantaneous among learners when the same learning content is the learning objective. Fast sharing leads to active learning, and learners receive instant feedback and tips. The active learning process in M-learning has increased the performance score from the fiftieth to the seventieth percentile and has reduced the dropout rate by 22 percent in technical fields [60]. M-learning gives potent portability to learners where books and notes can be replaced by tailored learning content. Some of the possibilities offered by M-learning, according to Fombona, Pascual-Sevillana, and González-Videgaray are greater access to information, along with transcendent innovations such as playful activities, learning in a group, iconic virtual and network of friends discussing on the same learning topic [61]. Chronologically, m-learning research can be separated into three phases: the first phase focuses on considering mobile devices properties and capabilities before using them as a learning tool; the second phase focuses on how mobile devices can be used to assist formal and informal learning and how mobile devices can be used for learning purpose outside of the classroom; the third phase in focus on mobility of the learner and how and what information to provide in diverse contexts [62]. In the second phase of m-learning, around 2005, many projects were completed worldwide, with each project targeting the effects of m-learning, such as increasing learners' engagement in learning activities, increasing learning motivation, and providing assistance according to the learning needs of learners. With m-learning projects' success and their wide acceptance among learners, m-learning now can be integrated with mainstream education and training.

The use of mobile devices is ubiquitous, with every field using it for their needs and requirements. In the education field, mobile devices are mainly used for classroom management, student-centered learning, group collaboration [63], distance learning [64], cross-contextual learning [65], lifelong learning [66], and self-learning [67]. Fuller Richard and Victoria Joynes listed some of the advantages of m-learning that encourage learners to learn in the workplace [68].

- m-learning helps learners improve their vocabulary, literacy, speaking, listening, and writing skills.
- Encourages learning to pinpoint the exact area where they need help and support.
- m-learning is relatively cheap as the cost of smartphones and other mobile devices are significantly less than computers.
- Decrease training costs where employees are separated by long-distance.
- Provide continuous personalized and situated learning support.
- Remove traditional classroom formalities for learning and give learners the freedom to engage in a comfortable and reluctant learning experience.
- Raises learners' self-confidence and self-esteem by allowing them to perform multitudes of different experiments.
- Learn without temporal and spatial constraints.

The continued improvement in mobile devices, wireless technologies, and cloud computing has enabled many potential implications on m-learning processes, such as cross-contextual learning, accessibility, appeal, and collaboration.

- Cross-Contextual learning: In cross-contextual learning, mobile devices provide help to learners in work settings according to what they have learned in formal educational settings [69]. Providing a cross-contextual learning experience is perhaps the biggest advantage of m-learning. Cross-contextual learning involves approaches in which learning in the workplace is facilitated and supported by reflective questions, formative assessments, and documentation of personal achievement in learning databases or portfolios [70]. The success of cross-contextual learning lies in integrating work-based learning experience with the formal learning experience and how effectively learners can apply what they have learned in the working environment.

- Accessibility: Due to mobile devices' portable nature, they have become a convenient tool for learners for providing them access to learning material regardless of their changing context [71]. The accessibility advantage is not limited to time and location contexts. The mobile devices used to access learning material using mobile networks are inexpensive and cheap compared to desktop and laptop computers.

- Appeal: The proliferation of mobile devices is sweeping the globe, and according to Cisco, m-learning in the U.S. market will grow from 7.98 billion in 2016 to 37.6 billion in 2020 [72]. According to Pew Research Centre statistics, 70 percent of U.S adults use smartphones of some kind [73]. Modern learners using smartphones prefer to consume a small chunk of learning material according to their choice rather than absorbing large-sized learning content served on desktop computers. In light of these facts, education organizations and business enterprises looking to improve their learners' or employers' skills will have to make learning/training programs more relevant, appealing, and effective. M-learning is more appealing as it gives greater learning flexibility where learners choose to select learning content in different formats – videos, audio, podcasts, and any other multimedia format. The lessons provided by m-learning are normally bite-sized, enabling learners to micro-learn while they are on the go. The ability of m-learners to choose their learning path to complete their lessons has increased content retention and learner engagement. Other factors that make m-learning more appealing include familiar learning devices, fewer learning restrictions and frustrations, engaging in collaborative learning, and selecting adaptive learning paths [74].

- Collaboration: Mobile collaborative learning includes using mobile devices such as smartphones, tablets, and personal digital assistants (PDA) in-class and out-of-class to share ideas and problems during learning. M-learning allows learners to collaborate with other learners and build knowledge collectively. M-learning allows students to keep up-to-date about current learning topics and eliminate the growth of unread text messages [75].

2.5 Mobile Cloud Computing (MCC)

MCC provides unrestricted processing and storage services to mobile applications (apps) based on the pay-as-you-use principle [76]. Multiple categories of cloud services are offered to users, including Infrastructure as a Service (IaaS) [77]; e.g., Windows Azure, Google Compute Engine, Rackspace, Amazon EC2, Platform as a Service (PaaS) [78]; e.g., AWS Elastic Beanstalk, Heroku, Google App Engine, IBM Bluemix, salesforce, Apache Stratos, Software as an Infrastructure (SaaS) [79]; e.g., Gmail, Google Drive, Evernote, Facebook.com, Office 365, Google Apps, Google Services, Backend as a Service (BaaS) [79]; e.g., Parse, Google Firebase real-time database, Firebase Storage, and Firestore. IaaS, as the name suggests, provides physical computing infrastructure, virtual machines, file-based storage, virtual disk image library, IP addresses, Firewalls, and virtual local area networks, etc. On the other hand, PaaS provides a computing platform such as web services, operating systems, programming language execution environment, databases, etc. The SaaS model user is given access to application software often referred to as "on-demand software". In a SaaS environment, a user does not have to worry about installing and executing application software rather than the service provider does it for the users. A user must use it through some client computer. In this research study, we use Google Firebase real-time databases and Firebase Storage for storing learners' feature instances, learners' authentications, database support, performance monitoring, and user analytics.

To facilitate learners in the m-learning process and overcome the limitations of previous studies, we developed an intelligent learning system that uses machine learning techniques to determine the exact needs and requirements of m-learners. An intelligent learning system will be discussed in detail in the upcoming chapters. Machine learning techniques and algorithms use learners' features stored in an online Firebase cloud database to determine their learning performance effects.

2.6 Context-Aware Learning

With the advancement in mobile devices and wireless technologies, mobile devices can both sense and react according to users' context and environment. Machine learning techniques integrated with mobile technologies can discover learners' learning behavior by mining and analyzing their context information. Learning context includes the learning features of m-learners. Learning features may consist of learners' interaction details, location, time, study duration, repetition rate, social interaction, background knowledge, performance, preferences, etc. Various experimental studies showed that mobile devices could be a handy tool in assisting learners in diverse learning settings.

In context-aware learning, mobile devices have information about the learning environment under which they make intelligent decisions and react according to circumstances, and situations [80]. It is essential to accurately categorize learners in context-aware learning and provide learning content and help according to their needs. In the computer science and educational community, context-aware learning initially was perceived as learning that considers changing locations, as Dev discussed [81]. In the past few years, context-aware learning considers many other features apart from just user location. Context-aware learning is now viewed as a learning process in which the learner is involved; thus, more sophisticated context models have been introduced and proposed. Context-aware applications now have many functionalities and features which include (a) provide a dynamic and personalized interface, (b) providing the needed information with high precision, (c) decreasing learner memory load, (d) make user interaction implicit and straightforward, (e) built smart learning environment.

Context-aware systems mainly target acquiring and knowing the exact context of computer users. The user context is made up of features or attributes generated during the user's interaction with the system. After acquiring user features, they are weighted and understood so that the perceived context matches the actual context. The context-aware system reacts and performs triggering actions according to user activities. User interactions and her/his activities are essential for intelligent applications; therefore, context awareness has been focused intensely on activity recognition research.

Aziz Hasanov and Teemu H. Laine classified the context-awareness process in the learning environment into three aspects: Context Acquisition, Context Modeling, Context entities [82]. Context acquisition deals with how to capture learners' current context using available technologies and tools. According to Perara et al., [83] there are three basics steps in context acquisition: (1) context can be captured by using sensors; (2) context can be derived from raw data through computation; (3) context can be taken directly by learners input. Context modeling involves computer algorithms and processes that can convert and understand necessary context information. Strang and

Linnhoff-Popien have classified context modeling approaches into six categories: Key-value pairs, Markup schemes, logic-based, graphical, object-based, and ontology-based [83]. In addition to these models, some researcher has also included a database (relational, object-oriented database, NoSQL) model to store and represent learners context information [83]. The database modeling approach is familiar and easy to use for storing and modeling context information. In a context-aware learning environment, context entities are those objects/items from where context information can be elicited. Verbert et al., [80] Li et al., [84] Gomez et al., [85] have established context entities taxonomy that categories context entities into five groups: learners (learner profile, interests, preferences, learning styles, cognitive abilities, social interaction information), technical entity (learner interface, operating system, CPU-occupancy, network bandwidth, memory and screen size, battery life), spatiotemporal (time, time duration, location), pedagogical (learning objectives, learning activities, learning tools, learning resources, pedagogical techniques), and environmental entity (noise level, temperature, humidity, luminosity).

Research studies carried out in the last decade reveal that mobile phones and PDAs are the most popular and common types of devices used in a context-aware environment. The popularity of mobile phones and PDAs in a context-aware learning environment is evident because they have high mobility and portability, allowing learners to traverse through context and switch contexts rapidly. In informal learning environments, mobility provides great affordance, such as in parks, museums, travel, science centers. Recently, mobile technologies had been a significant advancement, making their memory, processing, and other features equivalent to that of desktop computers. Nowadays, almost all mobile phone devices and PDAs provide an NFC or RFID reader module, allowing learners to interact with the surrounding environment easily. The utility and pervasiveness of mobile devices have made them an integral part of learners' daily lives.

2.7 Adaptive Learning

Sampson et al. define adaptive learning as a learning technology that provides a customized learning experience in which learners' unique needs and content are provided through just-in-time feedback, help, resources [86]. In traditional learning settings, the instructor is compelled to provide a one-size-fits-all learning experience without regarding the individual needs, performance, and preferences of different learners. Adaptive learning technology aims to support instructors in a traditional learning environment to provide the best possible learning experience for every individual learner. Using devices such as mobile devices and computers, adaptive learning techniques help instructors flex their teaching prowess to ensure that every learner progresses and learns. The power of adaptive learning techniques can be realized by the fact that they scale well to provide a customized learning experience to tens, hundreds, or thousands of learners at a time. Adaptive technology adapts to learners in several ways: i.e., designed adaptivity and algorithmic adaptivity. The graphical user interface (GUI) reacts to unique situations and learners' individual needs in designed adaptivity. GUI adaptation occurs based on adaptivity factors, including learner performance, interaction information, learning behavior, etc. Often, designed adaptivity react to the different situation according to the result of an "If this, then that" approach. Using the "If this, then that" approach, learners are provided with needed remediation, personalized learning content, and helps learners set their goals according to their learning path. In algorithmic adaptivity, computer algorithms try to learn what a learner "knows" and what a learner is expected to experience in the future. In algorithmic adaptivity, the computer can select the right path for a learner, which is according to learners' learning flow. Adaptive learning occurs dynamically without instructor interference. A common algorithm that falls in the algorithmic adaptivity category is Bayesian Knowledge Tracing (BKT) which calculates the rate at which learners perform learning activity [87]. Another known algorithmic adaptivity algorithm is Item Response Theory (IRT) which models the interaction of learners with isolated items [88].

There are many ways to analyze learner behavior and performance to personalize the learning content appropriately. Learner behavior and performance are analyzed based on adaptivity factors or features which include performance (how much a student knows about the topic and what s/he has performed in the past), background knowledge (prior or gained knowledge), preferences (likings), demographics, and other information sources. Adaptive technology can adapt the learning experience for the individual learner based on these adaptivity factors. Adaptivity features can make learning systems generate unique responses to learners. Based on a learner's learning journey, the learning system can move learners as planned, reroute them, provide help, give hints, etc.

Adaptivity or learning features make up the learner model. The learner model consists of information about learner goals, objectives, and knowledge. Every learner has a unique learner model that defines the adaptive learning path according to the learner features. Adaptive systems become more complicated in m-learning as the learner mostly does informal or casual learning in the m-learning process. Considering mobile devices' unique features and mobility, m-learning adaptive systems have become more sophisticated and specialized. They should focus on delivering adaptive learning content in bite-sized and presenting them simply and attractively. The adaptive learning system's primary focus is how to process and use the values of the learner's features during her/his learning process and how to make learning engaging, simple, and useful.

2.8 Machine Learning Techniques Used for Learner Modeling

Machine learning (ML) is a subfield in artificial intelligence used to develop analytical/learner models [32]. These analytical models automate the computer process, and human interference is not required. The fuel for ML algorithms is large datasets. When appropriately provided, ML algorithms learn iteratively from data and discover hidden patterns. ML algorithms work well where future predictions and estimations are required. ML techniques have been applied in a multitude of fields, including self-driving cars, stock exchange predictions, weather forecasting, email spam and malware filtering, online customer support, search engine results filtering, online fraud detection, product recommendations, voice recognition, image processing, marketing, economics, insurance, medical diagnosis, natural language processing, information retrieval, adaptive websites, object recognition, user behavior analytics and user behavior analytics in educational settings. The result of the ML-based learning model will be more accurate if the following conditions are true:

- The input data is available and is well understood by ML algorithms.
- The pattern between input and output features is well understood. It means initially for input data the output result is available and well understood.

- ML algorithms have been well trained on the input/output features pattern i.e. ML algorithms have enough experience on data patterns.

Recently, S.B. Kotsiants presented a novel case study, where he gathered demographics and learning data from written assignments for the ML regression method to estimate learners' future performance [89]. In this experimental study, S.B Kotsiants revealed that the education domain offers many interesting applications to ML. A learner model was developed based on the mapping between predictors features and target feature. Predictors features included basic demographic information such as gender, age, marital status, number of children, occupation, computer literacy, etc. Other predictor features included four face-to-face meetings with instructors and four written assignments. The target feature was the final examination score. Linear regression ML technique was compared to other popular and benchmark ML techniques such as Model Trees (MT, the counterpart of Decision Trees (DT)) [90], Artificial Neural Networks (ANN) [91], Locally Weighted Linear Regression (LWLR) [92] and Support Vector Machine (SVM) [93]. The comparison results revealed that Model Trees (MT) gives accurate results for constructing a software support learning tool. MT also gives better performance and is easily understandable.

A large amount of stored data in education learning environments such as online courses, online learning log files, demographics and academic information of students, registration/admission information, and so on can be useful for ML algorithms. ML algorithms use a reasoning approach to develop learner models that draw interaction between the learner and the learning system. Standard ML algorithms used for learner modeling include Artificial Neural Networks (ANN), K-Nearest Neighbors (K-NN), Bayesian network, Simple Linear Regression, Multiple Linear Regression, Fuzzy logic, logistic regression, Decision Trees (DT), Random Forest (RF), Apriori, Eclat and Naïve Bayes. All the ML algorithms mentioned above are called supervised ML algorithms. They are called supervised because input features have a label, and there is a relationship between input and output features. There are unsupervised ML algorithms such as K-Mean clustering, K-Median clustering, Hierarchical clustering, and Density-Based Spatial Clustering of Applications with Noise (DBSCAN), but unsupervised ML algorithms are rarely used in educational settings. In academic settings, there is a strong pattern in labeled features, and learners' input features have a relationship to output features. Unsupervised ML algorithms try to discover hidden patterns or clusters in unlabeled data where there is no association between input (predictors variables) and output (target variable) features.

ML techniques used for educational purposes can be divided into classification/ regression algorithms, sequential pattern analysis, association rules, clustering, and web information mining [94]. ML classification techniques group learners into different categories having similar learning behavior or attributes, i.e., grouping learners based on performance, misconceptions, hint-driven, or failure driven [95]; identifying learners with little interest in the study and recommending remedial actions to increase their learning motivations [96]. A research carried out by Hsia et al. [97] at a university in Taiwan used ML techniques to analyze the course's preference and course completion rate of learners. The Decision Tree technique was used to find learners' course preferences, and Link Analysis found the correlation between learners' profession and course category. In contrast, Decision Forest was used to finding the probability of learners completing the enrolled courses successfully. Similarly, Wang et al. used a decision tree algorithm for optimizing learning sequences in an adaptive learning system. The decision tree algorithm used a learner profile for discovering the most appropriate learning sequence for target learning content. The learner profile was created based on pre and post-tests and from a set of five learner features, namely: gender, cognitive style, personality type, learning style, and learners' grades from the previous semester.

Campbell proposed and develop a learner model using linear regression that demonstrated that learner SAT scores are insignificantly predictive of learner success [98]. With the inclusion of the LMS login feature, the predictive power of the model was tripled. The author demonstrated that the logins feature is directly proportional to the level of effort. The more log the learner performed, the more the effort s/he is undertaking. It was revealed that learners with a low or average SAT score could achieve success in final grade through above-average levels of effort. Macfadyen and Dawson tracked learners' LMS activities and collected 15 features that have a high correlation with learners' final grades [99]. The course that incorporated key features such as the total number of messages posted, the total number of email messages sent, the total number of quiz attempts, the total number of assignments completed, etc., explained 30% of the variation in the learner's final grades.

One of the famous and simple ML techniques used in educational settings is association rule mining [100]. Like other ML techniques, association rule mining tries to find hidden patterns in data. Association rule mining finds patterns by analyzing features that occur together and highly correlated features. Rules in association rule mining are not extracted from users' behavior or preference but instead finding relationships between elements in every distinct transaction. Various researchers have also used association rule mining techniques for educational purposes. Diagnosing learners' misconceptions, failed courses/dropout reasons, discovering learning problems, constructing course concept maps, improving online learning behavior, and promoting collaborative learning were key areas where the association rule mining technique was applied.

Numerous research studies have been carried out that used unsupervised clustering algorithms to group learners according to similar learning features. Anaya and Boticario used a clustering technique called Expectation-Maximization (EM) that group similar important features without knowing in advance about the structure of data [101]. In clustering algorithms, the features are group together according to their similarities by applying the Euclidean Metric. Clustering techniques grouped learners according to their online collaboration in distance learning settings. The results indicated that for successful collaborative learning, regular and frequent evaluation of team collaboration is necessary. The design of an online/distance learning environment must support a method that allows instructors to analyze learners' collaboration manually or automatically. However, when the number of learners exceeds a specific limit, regular and frequent collaboration analysis is a very daunting and time-consuming task for instructors. In those settings, the collaboration analysis should be performed using techniques that do not involve instructors. ML techniques are suitable for that scenario since they do not require human intervention and can be applied regularly and frequently in analyzing learner collaborations. Chen et al. used clustering algorithms, namely gray relational analysis (GRA), fuzzy association rule mining, k-mean clustering, and fuzzy inference, to identify learners' performance assessment rules using web-based learning portfolios of the individual learner. From these assessment rules, instructors can understand the factors that influence the learning performance of learners. Lin et al. applied two-stage clustering (Self Organizing Maps and K-means) on training data of automobile driving schools [102]. The result of clustering served as a reference point for future training courses. The author also developed an educational training prediction model using a back-propagation neural network. The prediction model was used to create a knowledge management system that assists the learner in automobile training courses.

Sequential pattern mining is the mining of frequently occurring patterns in data or mining of frequently occurring events. Sequential pattern mining techniques have been applied in many fields, including mining customer shopping sequences, web access patterns, weather predictions, etc. In educational settings, sequential pattern mining recommends a sequence of learning content/resources to learners to learn about a given topic. Perera et al. used sequential pattern mining to monitor the online learning activities of learners [103]. The monitored data/pattern is then used later to provide adaptive and needed guidance. The patterns were also extracted to characterize the learning behavior of different groups.

Web-mining techniques in m-learning/e-learning can be used to group similar World Wide Web (WWW) documents according to learners' target topics and similarities. Web-mining techniques can provide document summaries, inter-document patterns, hyperlink patterns among WWW documents, similarities, and relationships among different websites. Romero et al. proposed and implemented the architecture for a personalization system that recommends suitable weblinks/webpages to learners based on the web mining process [104]. The personalization system, namely AHA!, assisted instructors in carrying out the whole web mining process. With AHA! system, the instructor recommended a web page to the learner to visit next according to her/his interest and need. Similarly, Cohen and Nachmias provided a Web-log analysis tool for conducting a comprehensive analysis of the Web-logs and Web learning process [105]. Utilizing Web-log analysis measurements, the instructors analyzed and learned about learning content usage, learners' online interactions, learners' learning style, learners' attitude, the importance of website contents, and learning objects.

2.9 Use of deep learning techniques for educational purposes

Deep Learning (DL) is the most trending and discussed research area in artificial intelligence. DL is a subfield in ML that uses hidden layers in neural network models to represent data both at a granular and abstract level [106]. DL algorithms have small processing units inside hidden layers called neurons or perceptions that can apply linear and non-linear transformations to the input data. Different DL models and architecture have been proposed, developed, and successfully used in both supervised and unsupervised problems in the fields of image processing, natural language processing, handwriting recognition, self-driving cars, and computer vision, etc. DL algorithms form a hierarchy of multiple hidden layers where high layers drive data from lower layers. For example, in the image classification task, the DL model will represent the image at the input layer in pixels and assign a label to the image at the output layer. Between the input and output layers, there are hidden layers (transformation layers) that successively represent features at a more abstract level that are less sensitive to noise, lightings, and different image positions. The 'deep' in DL refers to multiple hidden layers that apply transformations to features, and at each layer, the features data is represented differently. The number of hidden layers in the DL model depends on the number of features and their instances. Research in the field of DL recommends that there should be at least one hidden layer in the DL model.

The concepts of DL were introduced and developed 30 years ago and some even long before. However, the implementation and achievements of DL have taken place in the last decade. Many reasons triggered the rise and success of DL algorithms, and it is agreed that the two major causes are the availability of a large amount of data and an increase in the processing power of Graphics Processing Units (GPU). In the first case, many data or big data allow DL algorithms to train and generalize well. In the second case, GPU allows DL algorithms to perform computational operations parallel to develop more comprehensive and deeper DL models. Another factor in the exponential emergence and development of DL has been the introduction of ML and DL libraries such as TensorFlow, PyTorch, Keras, and Theano. These ML and DL libraries help researchers focus on model architectures and structures rather than worrying about low-level implementation details and computer programming. Another advantage of the DL algorithm over ML algorithms is the avoidance of the features engineering process. In the feature engineering process, traditional ML algorithms select the most essential and representative features necessary for ML algorithms' operation to work, ignoring non-influential features. ML algorithms and researchers' process of features engineering is intimidating and time-consuming since the correct choice of features affects the overall performance and accuracy of models. The process of features engineering is carried out automatically by DL algorithms to discover the model's patterns at hand. Table 2.1 lists the famous DL architecture, comparisons with baseline algorithms, evaluation method, and performance score.

Table 2.1: DL algorithms in educational settings: DL algorithm/architecture used, baseline algorithms, evaluation method. The column performance indicates whether the DL algorithm outperformed baseline algorithms (Greater), under-performed (Lesser), or obtained the same results (Equal)

Reference	DL model architecture	Baseline ML model architecture	Evaluation method	Performance
Bo Guo et al., [107]	Students Performance Prediction Network (SPPN) based on sparse auto-encoder	Naïve Bayes, MLP, SVM	Accuracy	Greater
B Akram et al., [108]	LSTM	SVM, RF, Majority Class	Accuracy, Precision, Recall, F1	Greater
Wilton W.T. Fok et al., [109]	Deep Neural Networks	DT, Association Rules	Accuracy	Greater
Abhinav et al., [110]	MLP	K-NN, SVD	RMSE, MAE	Greater

Table 2.1 continued from previous page

Mayreen V. Amazona and Alexander A. Hernandez [111]	ANN	DT, Naïve Bayes	Accuracy	Greater
Alvarado et al., [112]	WE	N-grams	Precision, Recall, F-measure	Equal
Byung-Hak Kim et al., [113]	GritNet, bidirectional long short-term memory (BLSTM)	Standard Logistic regression	Performance	Greater
Fei and Yeung et al., [114]	RNN, LSTM	SVM, LogReg, IOHMM	Accuracy	Greater
Samuel-Soma et al., [115]	Ensemble Techniques	NB, DT, K-NN, Disc, PWC	Accuracy, Performance	Greater
Mirza Mohtashim Alam., [116]	Artificial Neural Network	Random Forest, Chi2	Classification, Accuracy	Equal
Khajah et al., [117]	LSTM	BKT	Accuracy	Equal

Table 2.1 continued from previous page

Xiaofeng Ma et al., [118]	DNN	DT, SVM	Prediction, Accuracy	Lesser
Lalwani and Agrawal et al., [119]	LSTM	PFA, BKT	Accuracy	Equal
El Fouki et al., [120]	DNN	PCA	Prediction	-
S Abidi et al., [121]	DL	Generalized Linear Model (GLM), Logistic Regression (LR), Decision Tree (DT), Random Forest (RF), and Gradient Boosted Trees (XGBoost)	Leaners confusion Prediction	Greater for RF, GLM, XGBoost, and DL
Hadullo et al., [122]	MLP	-	Performance prediction factors	-

Table 2.1 continued from previous page

W. Wang et al., [123]	CNN, RNN	SVM, LogReg, DT, AdaBoost, GTB, RF, GNB	Accuracy, Precision, Recall, F-measure	Equal
I. Ndukwe et al., [124]	DNN	-	Classification	-
Whitehill et al., [125]	FNN	-	Accuracy	-
Y Chai et al., [126]	ANN		Accuracy	Greater
Yeung and Yeung [127]	LSTM	-	Accuracy	-
S Yanchao et al., [128]	CNN	ANN	Prediction, Features reconstruction	Greater
E Tanuar et al., [129]	ANN	Linear Model, DT	Accuracy	Greater
Yang et al., [130]	LSTM	-	MSE	-
DYUTI ISLAM [131]	MLP	RF, SVM,	Accuracy, Precision, Recall, F-measure	Greater
Xizhe Wang [132]	Convolutional GRU	RF, SVM, BPNN, RNN, LSTM,	Accuracy, F1-Score	Greater

Table 2.1 continued from previous page

Amjad Abu Saa [133]	NN	RF	Prediction accuracy	Lesser
Sales et al., [134]	LSTM	-	-	-
Sharma et al., [135]	CNN, AlexNet, VGG16, LSTM	SVM, HMM	Accuracy	Greater

2.10 Neural Networks (NNs)

The following section describes the foundation of NNs, their architecture types, training process, hypermeter tuning, and elements of NNs.

NNs is a set of algorithms named loosely after the working of neurons network in the human brain. NNs are designed and developed mainly for predictions, pattern recognition, and the area where a lot of information processing is involved [136, 137]. Like ML algorithms, NNs algorithms are fundamentally divided into two forms. Supervised and Unsupervised NNs. Supervised NNs are mainly used for regression, classifications, computer vision, and time series analysis tasks where the NN is first trained using input and output features. Then predictions are made on unseen data in the future. Unsupervised NN is primarily used for feature detection and recommendation tasks where hidden patterns need to be discovered in raw data. Examples of supervised NNs algorithms include Artificial Neural Networks (ANN), Convolutional Neural Networks (CNN), and Recurrent Neural Networks (RNN). Examples of unsupervised NN algorithms include Self-Organizing Maps (SOM), Deep Boltzmann Machines (DBM), and Autoencoders.

NN process input data through neurons or machine perception, and the result of processing can be labeled data, classified data, clustered data. NNs can process all types of real-world data such as images, texts, sound, or time series. But all types of data must be translated into numerical forms or contained in vectors for NN algorithms to recognize patterns inside it. A simple structure of a neuron or a node in NN is depicted in figure 2.1. A neuron's input is the features instances or feature data (x0, x1, x2,...xn) and the weights of synapses or edges (w0, w1, w2,...wn). A neuron performs two operations in the NN model. First, it multiplies input feature data with corresponding weights, and then the activation function is applied to the multiplication result. Activation functions will be discussed later in this chapter. The input to a neuron can be the output of another neuron in the same NN model. The input data also consists of a bias value which is a constant value added to the input values before the activation function is performed. The weights values represent the importance of features data inside NN or the significance of neurons inside NN. Every neuron has an output value represented by (α) and is generally calculated by equation 2.1. Output value can be continuous (price, score), binary (yes/no), or categorical (below-average performance, average performance, good performance).

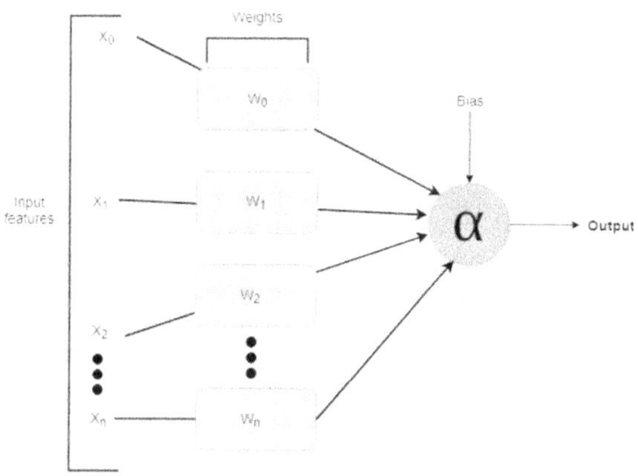

Figure 2.1: Artificial neuron with inputs, output, activation, and edges weights

$$\alpha = f\left(\sum_{i=0}^{n} X_i W_i\right) \tag{2.1}$$

f in the above equation is the activation function applied to the result of summation operation performed after input values are multiplied by corresponding weights. The activation function allows NN to evaluate the non-linear relationship between input/output data and provide standardization that affects the neuron output (bound the output result between 0 and 1). The primary purpose of the activation function is to convert the input data into output data. The output data generated by one neuron can act as input data to another neuron. The extensively used activation functions are ReLU (Rectified Linear Unit), sigmoid, Softmax, and tanh (hyperbolic tangent). The neurons inside the NN are connected, and the activation result of one neuron can upsurge or constrain the adjacent neuron's activation state.

Figure 2.2 shows the basic structure of NN architecture. It consists of an input layer, a hidden layer, and the output layer. The input layer delivers feature instances to NN, whereas the prediction result is generated at the output layer. Initially, simpler functions are applied to data at the hidden layer, but by cascading simple functions in multiple hidden layers, NN can perform complex functions. The more the number of hidden layers, the more complex operation NN can perform. There can be hundreds or even thousands of hidden layers in NN architecture, depending on the size of data and complexity of the operation. Once the learning process of NN-based models is complete, they can then make predictions automatically on new and unseen data. Different activation functions are applied at the output layer depending on the nature of the problem: for binary classification, where the output result can only be 0 or 1, threshold or sigmoid activation function can be used; for multiple classification or categorical problem, where the output can be more than two values, Softmax activation function is used. Lastly, the linear activation function is used where the output continues or we have regression problems. ReLU (Rectified Linear Units) is the most common activation function used in the hidden layers. ReLU returns the number itself for positive numbers, whereas, for negative numbers, it returns 0. The type and structure of hidden layers determine different NN architectures, such as ANN, RNN, CNN or LSTM, etc. The following equation 2.2 shows the basic functionality of ReLU.

$$f(x) = \begin{cases} 0, x < 0 \\ x, x >= 0 \end{cases} \tag{2.2}$$

2.10.1 Training NNs

In the training phase, NN input records have a known output class. There is a correct mapping between input record and output class, which is why it is called supervised

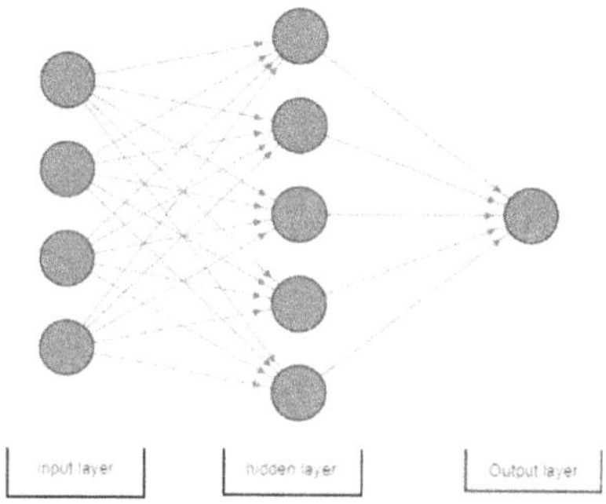

Figure 2.2: Basic architecture of Neural Networks having input, hidden, and output layers

learning. An input record maps into one of many output classes or categories. A neuron in the output layer will generate 1 to identify a particular class if input data correctly maps to it, and the rest of the neurons will generate an output of 0. The more the input records, the better the NN will be trained. NN compares the predicted value to the correct output during the training process to know how close the predicted value is to the correct output. The difference between the predicted value and the correct value is called the error term. NN tries to minimize the error term during the training process, thus reducing the difference between the predicted value and correct value. The error term is backpropagated in the NN, and the weights of hidden layer edges are adjusted so that in the next iteration, the output value will be closer to the actual correct value. For example, NN may accept input as an image (e.g., a cat) and assign an output label to it (e.g., the cat label). The more a NN is trained on several different cat images, the better the NN will learn and predict. As mentioned earlier, the NN maps input data to target output using the hierarchy of multiple hidden layers. At early hidden layers, the NN learns something simple at a finer granularity. Each hidden layer makes changes to feature instances and adds more complex information to them. As the feature instances traverse through the NN, more complex information is added to them at each hidden layer. How much information the NN has learned at each hidden layer about the feature instances is stored as weights in hidden layers corresponding edges. As feature instances pass through the NN, the level of granularity increases, and learning becomes coarser. For a NN to learn, it is important to find the most suitable edges' weights that best represent the mapping between input data and the corresponding target output. Initially, during the NN training process, the weights on edges can be assigned randomly (small numbers closer to 0) or follow some initialization strategy. With machine learning libraries like TensorFlow, Keras, Theano, etc., the researchers do not have to worry about assigning weights to neuron edges. The TensorFlow machine learning library provides Python Dense class that is usually used to assign weights to neurons' edges.

Finding the most suitable weights for neuron edges during the NN training process can be the most time-consuming and challenging task. Initially, millions of weight values and NN have to select the correct value for each edge during the training process. The most suitable edge weights correctly map input features to the target output. To accurately predict NN's output, it is necessary to quantify how close the obtained output is to the actual output. The NN loss function carries out measuring the difference between the obtained output and actual output. The results of the loss function indicate how well the NN has learned on specified examples. Mean Squared Error (MSE) is a typical example of a loss function that calculates the Mean of a squared error made by the NN over the input features instances. The training process's goal is to minimize the difference between predicted output and actual output or to get a minimum value for loss function. The error calculated by the loss function is backpropagated through the NN, and the weights of neuron edges are adjusted accordingly. The updated weights on neuron edges guarantee that error will be reduced in the next training phase. The amount by which neuron edge weights are updated is determined by a parameter called the learning rate. The weights updation is generally carried by a nonlinear optimization technique called stochastic gradient descent (SGD). The NN computes the derivative of the generated error for weights and updates them such that the error in the next iteration will be decreased. SGD is more impressive when we have a large amount of input data, and we are trying to find optimal weights that will generate a minimum error in the output result. A NN completes one training cycle (epoch) when the whole features instances are feedforward through the NN. After the number of epochs, the NN usually covers a state where the difference between predicted output and target output is minimal. When the error reaches the minimum value, the NN is considered to have learned the target function (mapping between input instances and target output).

2.10.2 Types of NN Architectures

There are different types of NN architectures, and their use depends on the kind of input data (video, audio, image, text, numbers, etc.). One type of NN architecture may be useful in processing images, whereas another architecture may be good in processing text or time-series data. The number of NN architectures used in research and development is extensive and diverse. This section will describe the most common NN architectures, their task, and their applications in learning technologies/educational settings. Table 2.1 describe the application of deep learning-based NN architectures used in learning technologies, the type of architecture used and implemented, comparison with the baseline ML algorithms, the evaluation method used to compare DL based NN algorithms with ML algorithms, and the performance (greater, lesser, equal) of DL algorithms in comparison with baseline ML algorithms.

The deep neural network (DNN) architectures include Artificial Neural Network (ANN) [138] with multiple hidden layers, Convolutional Neural Networks (CNN) [139], Multilayer Perceptron (MLP), Recurrent Neural Networks (RNN) [140], Long Short-term Memory (LSTM) [141], Self-Organizing Maps (SOMs) [40], Deep Boltzmann Machine (DBM) [142], Deep Auto Encoder (DAC) [143], Word Embeddings (WE), Bidirectional LSTM (BLSTM) [144], CNN variants (VGG16 and AlexNet) [145], and Memory Networks (MN) [146]. In table 2.1 the baseline ML algorithms include Naïve Bayes, Support Vector Machine (SVM), Random Forest (RF), Linear Regression (LinReg), Decision Trees (DT), Association Rules, Standard Logistic Regression (LR), Bayesian Knowledge Tracing (BKT), Principal Component Analysis (PCA), Generalized Linear Model (GLM), Gradient Boosted Trees (XGBoost), AdaBoost, Gradient Tree Boosting (GTB), Gaussian Naïve Bayes (GNB), Hidden Markov Model (HMM), and Back-Propagation Neural Networks (BPNN). The evaluation method includes algorithms accuracy, Precision, Recall, F1, Mean Absolute Error (MAE), Root Mean Square Error (RMSE), F-measure, classification, and Mean Square Error (MSE). The last column in the table DL indicates whether deep learning algorithms with baseline ML algorithms were greater, lesser, or equal. Equal performance indicates that in some evaluations, DL algorithms showed less performance, and in others, they showed greater performance compared to ML algorithms. Lastly, the symbol "-" indicates that DL algorithms were not compared to ML algorithms; instead, they just compared different DL architectures or hypermeters used in different DL architectures. Next, a description of different DL algorithms is detailed, along with their applications in educational settings.

2.11 Artificial Neural Networks (ANN)

A simple ANN consists of one input and one output layer. The input data in this type of ANN is passed directly from the input layer to the output layer via a series of weights. The weights represent the strength of edges that connect input layer neurons to output layer neurons. Each neuron at the output layer calculates the sum of the product of weights and input data. After the sum of product operation, the activation function is applied to its result. If the activation function results are above some threshold value, the output is generated; otherwise, the neuron stores the deactivated value, and the output is not generated. ANNs having one input and one output layer are only capable of learning linear patterns. By introducing multiple hidden layers, ANNs become more flexible and powerful. With hidden layers, ANNs are capable of learning both linear and non-learner patterns. In ANN architecture, there is no cycle, and information always propagates in one direction, i.e., from the input layer to hidden layers and then to the output layer. That's why ANN is also called a feedforward neural network or multilayers perceptron. An architecture of ANN with two hidden layers represented in figure 2.3.

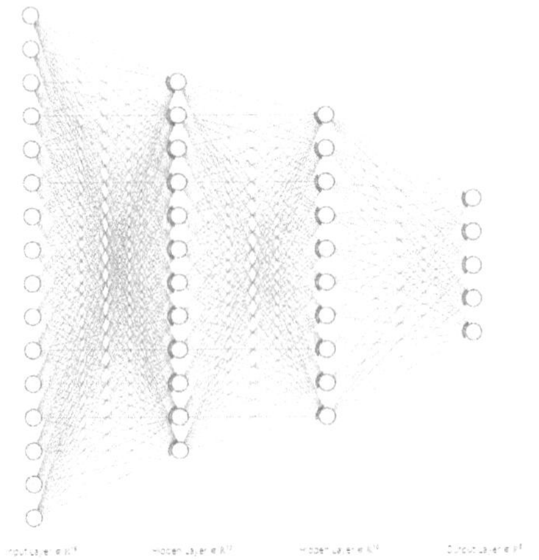

Figure 2.3: ANN architecture with one input, two hidden, and one output layer.

The original goal of ANN was to think, process, and solve problems in the same way as the human brain would do, but with time attention moved from performing a particular task to diverse tasks, leading to deviation from the field of biology. Currently, ANN applies to many areas where classical ML algorithms have been used. Still, major successes came in the area of image processing [147], speech recognition [148], computer vision [149], social networking advertisement, and filtering [150], and medical diagnosis, etc. ANNs offer good results and performance where the number of layers, input data, and neurons is large enough. Mostly ANN is used for supervised learning tasks where initially input features have known output results.

In educational settings, ANN has been applied for recommending suitable learning content to learners based on their performance, and preferences [151], [152]. ANN has also been applied in predicting learners' performance based on their learning behavior, and background knowledge [153].

2.12 Deep Autoencoders

Autoencoders are unsupervised, directed types of neural networks, where like feed-forward neural networks, the data propagates only in one direction, i.e., from input to output layer. In autoencoders, the number of neurons at the input layer is always the same as the number of neurons at the output layer. The autoencoder aims to encode the data at the input layer and then decode the same input data at the output layer. For encoding, the autoencoders need one or several hidden layers depending on the types of input data and types of compression techniques. The autoencoder aims to produce the same data at the output layer received at the input layer. That's why they fall into the category of unsupervised or self-supervised DL algorithms. They are called self-supervised because they are comparing output results to the input data. They are also called unsupervised DL algorithms as no label data is required at the input layer. The goal is to reconstruct the input data at the output layer rather than performing prediction on unseen data.

Autoencoders are commonly used for features detection, compression, recommendations, etc. [154, 155, 156]. For example, the neurons at the autoencoder's hidden layer can encode or represent some features, i.e., learner performance, preference, background knowledge from the data that is accepted at the input layer. The encoded features can then be used in the future for learners' guidance and feedback. Autoencoders (and their variants sparse, deep, contractive, stacked, denoising) are designed to compress or encode data and, subsequently, learn from that compressed data. Another autoencoder application (sparse autoencoder) is to pre-train a deep neural network: a stacked autoencoder is trained in an unsupervised fashion, and the weights of edges are obtained [157]. These weights are then used in deep neural networks (the same configuration in terms of hidden layers and neurons in each hidden layer) as a better choice to initialize the neural network rather than initialize them with random weights. After assigning weights to hidden layers, the deep neural network is further trained in a supervised learning fashion, and the parameters are also fine-tuned. Figure 2.4 shows a simple autoencoder with one input, one hidden, and one output layer.

Focusing on the use of autoencoders in learning technologies, the work done by [107] used a sparse autoencoder for the task of predicting learners' performance. Another work done by [42] used autoencoder to automatically extract and analyze learners' features from the learner model. The learner model was developed on the data extracted from learner interaction with the online learning system. The extracted features were further used to classify learners into different learning groups based on their preferences, interaction frequencies, and performance.

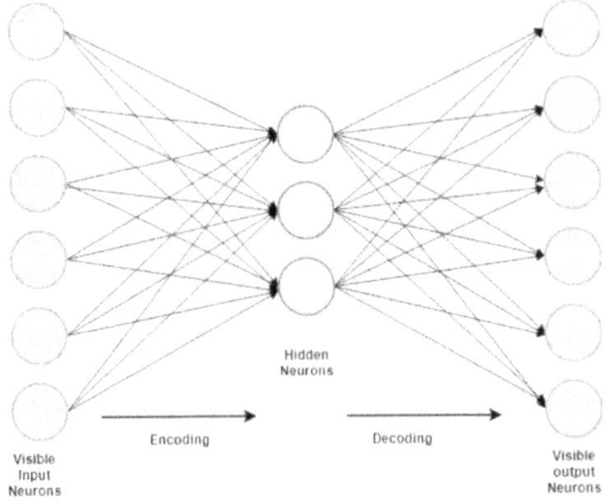

Figure 2.4: Autoencoder in the encoding-decoding process where the input data is reassembled at the output layer

2.13 Convolutional Neural Networks (CNNs)

CNN's models are mainly famous for processing images and videos. Like ANNs, CNNs are multilayer networks developed particularly for image processing applications. The CNN model accepts the input image at an abstract level, processes it, and classifies it into one of many categories (e.g., cat image). The CNN model performs four steps at a detailed level: convolution, max pooling, flattening, full connection for image classification. In the first step, CNN recognizes simple features in images, e.g., pixel density, edges, color combinations, and in the last step, it combines those low-level features into higher-level abstraction, e.g., forms an image, recognizes faces, etc. Though developed for processing images, CNNs are analogous to ANN in many aspects: CNNs are made up of neurons where weights and bias have to be learned, neurons accept input values, multiply it by weight, and performs activation function of the sum of products, and there is a loss function at the output layer that compares predicted result to actual result. At the convolution step, CNN develops a feature map by performing dot operation on the input image matrix (2-dimensional array) and feature detector matrix. In the convolution step, CNN's main objective is to reduce the image's size and represent it in a concise feature map retaining essential features of the actual input image. Several feature detector matrices are used (depending upon the situation) to generate feature maps from the input matrix. The features maps could be hundreds or thousands in numbers depending on the convolution layers' requirement and how much features information needs to be preserved. Figure 2.5 shows the convolution step where the features map is obtained by multiplying the input image matrix with the feature detector matrix. Figure 2.6 shows a pool of feature maps generated at the convolution step.

To increase non-linearity and filter important feature information, CNN also applies the ReLU activation function on each value of feature maps in the convolution step. ReLU activation function is used to feature maps, and non-linearity is increased because images themselves are highly non-linear, especially when CNN is trying to recognize different images next to each other.

In the second step, CNN applies the Max Pooling process to features maps. Max Pooling is a sample-based discretization method where an input image or hidden layer output matrix is down-sample, thus reducing its dimensionality and allowing assump-

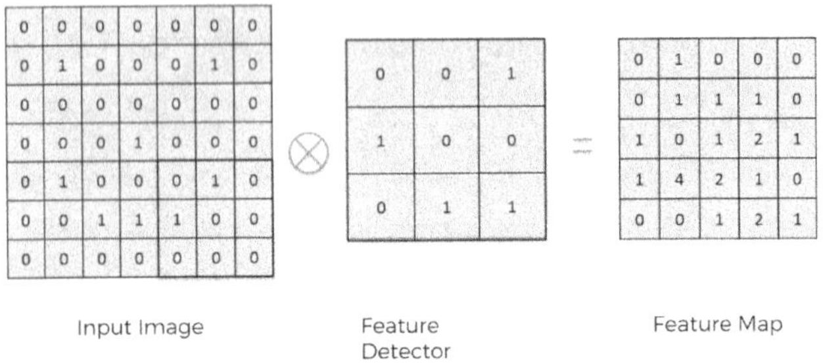

Figure 2.5: A single feature map is generated by multiplying the input image matrix with the feature detector matrix

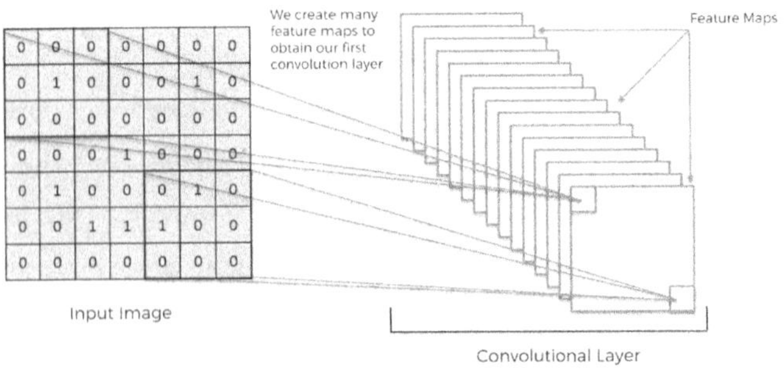

Figure 2.6: Stack of features maps generated at the convolution layer

Figure 2.7: Max pooling step

tions to be made about features in a pooled feature map as shown in figure 2.7. Max Pooling allows CNN to neglect trivial features information and preserve important features information in a pooled feature map. In this technique, maximum values from a feature map are extracted and recorded into another smaller matrix called pooled feature map.

In the third step, CNN took the pooled feature map and flattened it into one column. Each row from the pooled feature map is extracted and put inside one huge vector of input data. The features information is ready to be inserted into the neural network for further feature processing. The flattening step is shown in figure 2.8.

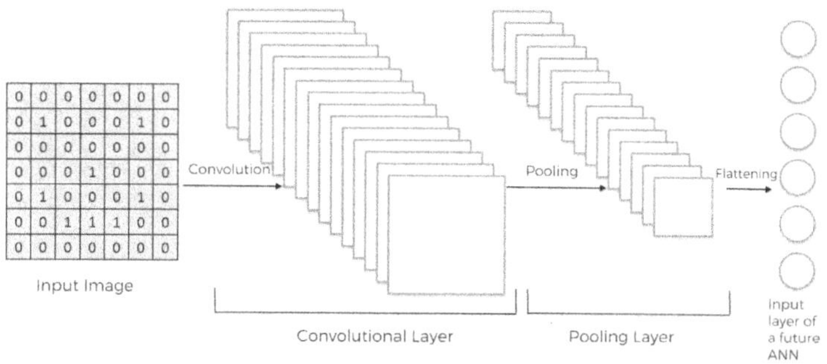

Figure 2.8: Flattening process where features are put in one huge input vector column

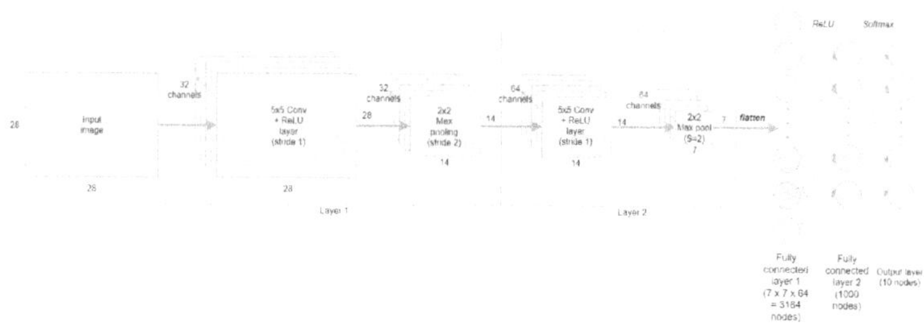

Figure 2.9: CNN connected with a fully connected ANN

In the fourth step, ANN is added to CNN as depicted in Figure 2.9. In this type of ANN, the hidden layers must be fully connected where each input and output neuron is connected with the neuron of hidden layers. ANN's primary purpose in CNN is to combine detailed features with more abstract features to predict the classes (e.g., images) accurately. The result of the flattening step produces features encoded into one colossal vector. The features in vector could be used to make predictions, but ANN architectures are developed to accept low-level features, combining them and coming with better prediction results. In simple words, ANN, when integrated with CNN, leverages the prediction accuracy. The use of fully connected hidden layers and convolution, max pooling, and the flattening process has facilitated the emergence of new CNN applications, particularly in image processing, game playing, natural language processing, video recognition, etc. The CNNs have greater accuracy and require fewer traditional

ANN parameters in pattern recognition tasks such as natural language processing and image processing.

On the other hand, CNN experiences high computation costs and requires a large amount of training data in properly initializing the network according to the problem addressed. In learning technologies, CNNs have been used in detecting undesirable learners' behavior using VGG16, and AlexNet architectures for analyzing learners' recorded videos [158, 159]. The CNNs have also been used in text classification tasks [160] and predicting learners' dropout [161].

2.14 Recurrent Neural Networks (RNNs)

RNN based algorithms are one of the most advanced algorithms in the world of supervised deep learning. In RNNs, neurons make a decision based on current input data and what they have observed previously. The decision that RNN has taken in time t-1 will affect the decision that it will take at time t. In contrast to ANN, RNN has two sources of information to decide on, i.e., the present input and the recent past output. Both present and the recent past output combine to determine how neurons will respond to new data, much like humans make decisions in their daily life. A distinguishing feature of RNNs is that they have a feedback loop that connects them to their past decisions. RNNs ingest their new input data and the past recent decision to come with a new output decision. The feedback loop allows RNNs to store or persist information for a short time, something which ANN cannot do.

The structure of RNNs is similar to ANN except that the network has multiple copies of themselves, as shown in figure 2.10. The information is preserved in the short-term when a network passes a message to its similar successor network. The structure of RNNs makes them the perfect tool for dealing with text, time-series data, and lists. Thus the common application of RNNs is image processing, sentiment analysis, video analysis, speech recognition, language modeling, natural language processing, and modeling text. Like other neural networks, RNNs also have some drawbacks, one of which is vanishing gradients. During the backpropagation phase, the magnitude of values used to update the neural network weights gets exponentially small, resulting in slower learning of weights in the lower layers of RNN.

Due to the vanishing gradient problem, the RNN ceases to learn more because if we

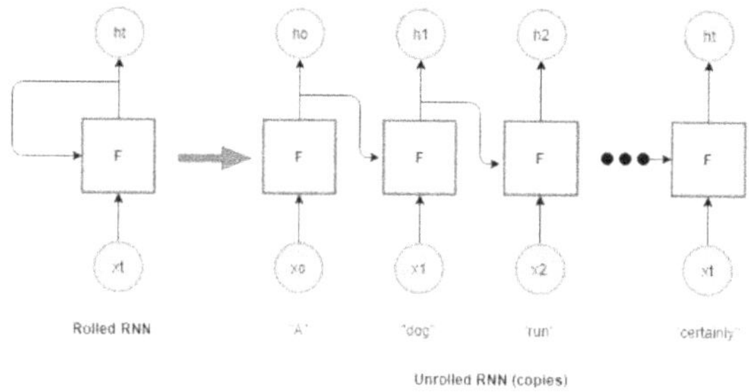

Figure 2.10: RNN with multiple copies used for data persistence

do not know the gradient values, we cannot adjust the weight to decrease the generated error at the output layer, as shown in Figure 2. 11. Another disadvantage of RNNs is that they require high computation costs to train, test, and run the models. In learning settings, RNNs have been used in predicting learners' dropouts [162, 163] predicting learners' performance for learning gain predictions [164], and proficiency assessment

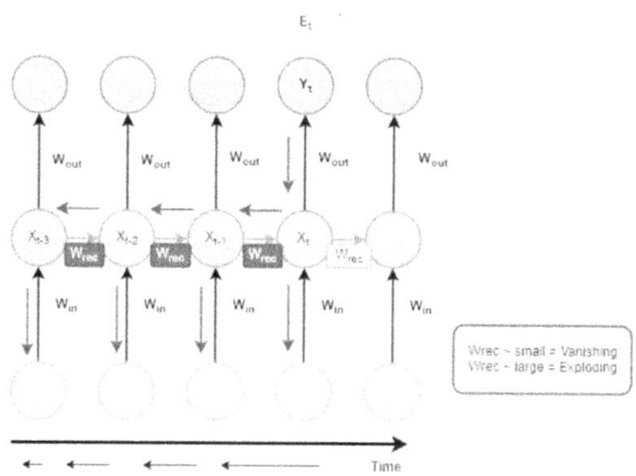

Figure 2.11: The vanishing gradient problem in RNN

Due to the vanishing gradient problem, the RNN ceases to learn more because if we do not know the gradient values, we cannot adjust the weight in the direction that will decrease the generated error at the output layer figure 2.11. Another disadvantage of RNNs is that they require high computation costs to train, test, and run the models. In learning settings, RNNs have been used in predicting learners' dropouts [68], [98] predicting learners' performance for learning gain predictions [99], and proficiency assessment [100].

For training, RNNs rely on the modified form of backpropagation technique called backpropagation through time or BBTT. Time in the BBTT technique is expressed by an ordered series of calculations linking a one-time step to another, which all backpropagation needs to perform their task. Long short-term memory (LSTM) is a type of RNN architecture (discussed in the next section) that removes many problems (including vanishing gradients) of RNNs.

2.15 Long Short-Term Memory Networks (LSTMs)

LSTMs are particular types of RNNs that work for many problems and tasks much better than the standard RNN version [101]. Almost all the problems/areas addressed by RNNs are also be addressed by LSTMs, such as image captioning, machine translation, speech recognition, image processing, sentiment analysis, etc., with much efficiency and accuracy. LSTMs emerged as RNNs could hold/remember long-term dependencies. RNNs can persist short-term information, but RNNs become powerless to learn to connect information when the gap between relevant information grows. For instance, sometimes, the machine needs recent information to decide on the current task. A language model could be in a state where it predicts the next word based on the previous word in a sentence. If the language model is trying to predict the last word in the sentence "children are playing in the playground...", the language model does not need any further or past context to predict the last word. For the language model, it is pretty clear that the last word would be a playground. In such a case where the gap between the past information and the place where it is needed (current context) is very small, RNNs can learn and use past information. In machine learning, especially in language modeling, sentiment analysis, etc., there are cases where more contextual information is needed to process current information. For example, if a language model is trying to predict the last word in the text, "I grew up in Saudi Arabia... I can easily and fluently speak Arabic". Recent context information suggests that the next word that the language model is trying to predict in Arabic, i.e., the name of the language, but if we try to narrow down which language, we need more considerable context information and the name of the countries where Arabic is spoken. It is utterly probable that the gap between the current task and the required information to be huge. If the gaps grow large, RNNs cannot learn to connect information. This is a significant shortcoming of RNNs, which LSTMs try to address. LSTMs are capable of learning long-term dependencies. As compared to RNNs, they work extremely well in many application areas and are now broadly used.

The architecture of LSTMs is designed so that they can persist information for long periods by default, as shown in figure 2.12. Like RNNs, LSTMs have a similar chain-like structure, but hidden layers operations (repeating groups) have a different structure. Inside the one repeating group, there are four neural network layers, interacting in an extraordinary way that enables LSTMs to store long-term dependencies. LSTMs have a memory/cell pipeline that goes through each repeating module carrying input vector information. The memory/cell pipeline is the straight line that passes through the LSTM repeating modules. The input vector information can sometimes be removed, or sometimes things might be added to it. The error is backpropagated through this memory/cell pipeline during the backpropagation step, preventing LSTMs from falling into the vanishing gradient problem. Each LSTM repeating module has three inputs and three outputs, i.e., the inputs are memory cell, input data, and the previous repeating module's output. The outputs are memory cell, output data to the world, and output data the next repeating module.

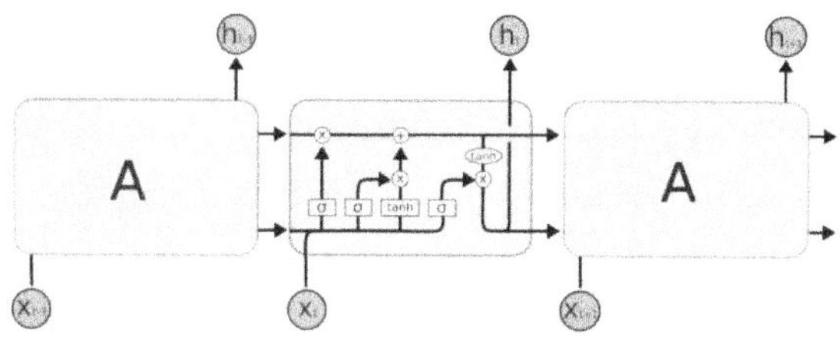

Figure 2.12: LSTM architecture with three repeating modules

In educational settings, LSTMs have been used and applied in a multitude of tasks, including detecting undesirable learners' behavior that causes learners dropout [102], predicting learners performance [103], modeling learners behavior in the online learning platform [75], generating recommendations and help [104], predicting engagement intensity [75] and automating essay scoring [105].

A variant of LSTMs called bidirectional LSTMs, or BLSTMs, has also been employed in educational settings to predict learners' performance. The difference between conventional LSTMs and BLSTMs is that the former preserve information from the past whereas the latter runs input in two directions: one from the past to the future and the other from the future to the past, storing information of the future in the backward path.

We used ANN for M-learners classification because m-learners classification based on their learning features is a supervised learning task. For classification and prediction problems, ANN are usually preferred deep learning algorithms [165]. Long Short-Term Memory (LSTMs) are used to solve problems related to time series analysis, speech recognition, handwriting recognition, human action recognition, and time-series predictions [166]. Convolution Neural Networks (CNNs) targets areas such as facial recognition, document analysis, historical and environmental collections, understanding climates, advertising, and image processing [167]. Similarly, Recurrent Neural Networks (RNNs) are used for text summarization, text auto-fill, language translation, call center analysis, and digital asset management in marketing [168]. Self-organizing maps are focus on project prioritization and selection, Seismic facies analysis for oil and gas exploration, failure mode and effects analysis, and creation of artwork [40]. Lastly, Deep Boltzmann machines target dimensionality reduction, recommender systems, and topic modeling [169].

This chapter discussed the application of various ML and DL techniques used in online learning environments. Several learning technologies were examined to understand how much they can help instructors and learners manage their teaching and learning goals. We discussed distance learning, e-learning, M-learning, Mobile cloud computing, context-aware learning, adaptive learning, machine learning, and deep learning. We also compared various DL and ML algorithms in terms of performance used in educational settings. Last, we briefly discussed popular DL algorithms and how they were applied in online learning environments.

PROPOSED M-LEARNING FRAMEWORK

This chapter is further divided into two sections. The first section is about the mobile learning system (learnit) framework, development, testing, and deployment. The development of the mobile learning system was necessary for feature generation and gathering. Section two describes how learners used learnit, the types of learning features, and how learning features were collected from it. Learning features were gathered from learner's interaction with learnit through mobile devices.

3.1 Mobile learning system framework

Figure 3.1 presents the M-learning system's overall framework containing four phases. The details about each phase are discussed in the below sections and chapter 4.

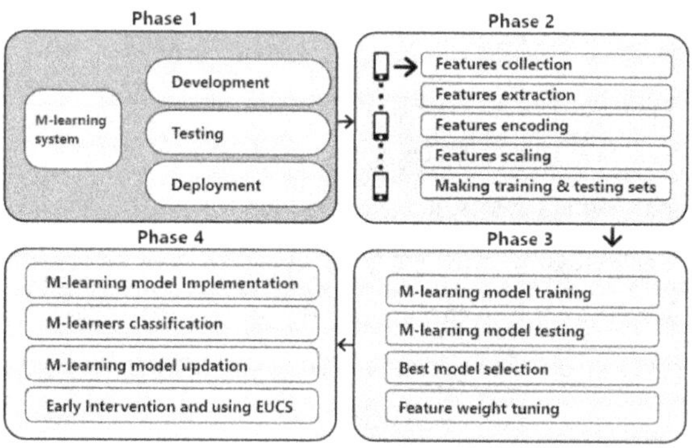

Figure 3.1: M-learning system framework is divided into four phases. Phase 1 is about M-learning system development, testing, and deployment. In phase 2, learning features are collected, extracted, encoded, scaled, and split into training and testing sets. In phase 3, the M-learning model is trained, tested, and integrated into the M-learning system for providing adaptive support to M-learners. Phase 4 M-learning model implementation, M-learners classification, M-learning model updation, performing the early intervention, and measuring M-learners satisfaction using EUCS.

3.1.1 Learnit app development

We used Android Studio Integrated Development Environment (IDE) for the development of the learnit app. Extendable Markup Language (XML) was used for designing the interface of the learnit app, whereas to add functionality to the design components, JAVA language was used. Details about learnit app interaction with the M-learners are discussed in the next sections. The backend used to store M-learners features was Google Firebase Realtime Database, which saves data in JAVA Script Object Notations (JSON) tree.

3.1.2 Mobile learning system (Learnit) interaction

Learnit was developed for both instructors and learners to enhance their teaching and learning activities. By using learnit, instructors can add courses, course content, and quizzes on Firebase cloud. The course content that instructors can add was in videos and text (PDF documents). The quizzes were of multiple-choice questions (MCQs) where one question has one correct answer. In the deployment phase, we added three courses, namely Computer Basics, C++, and JAVA. Each course had three modules, and a quiz or checkpoint was conducted after completing each module. We also administered a final quiz where the final performance of learners was evaluated.

For attracting learner's attention towards learnit, we kept the average video length of 5 to 8 minutes. This was because concise learning content, when provided in small chunks on mobile devices, attracts more learners as learners are on the move and are more vulnerable to external environment distractions and noises. In an environment where there is a lot of distraction and noise, bite-sized learning content is perfect for learning as learners cannot hold their attention for large learning content on mobile devices for a longer time. Each course quiz was 5 points, whereas the final quiz was 10 points. Later, the course quizzes and final quiz points were converted into 20 points for better analysis and how much they have effects on the final performance. Figure 3.2a, Figure 3.2b, Figure 3.2c, Figure 3.2d, Figure 3.2e, Figure 3.2f, shows the interaction of learners with different activities of the learnit app whereas Figure 3.3a, Figure 3.3b, Figure 3.3c, Figure 3.3d, Figure 3.3e, Figure 3.3f shows learning, problem posting and quiz taking activities. For this study, three courses (computer basics, C++ programming, JAVA programming) were offered to learners. While using learnit, a learner can select any course s/he is interested in. Each course is further divided into three learning modules. Each learning module consisted of three learning topics. After the completion of each module, a quiz activity was offered to learners to evaluate their learning.

A final quiz/checkpoint was administered after completing three courses to evaluate the learners' overall learning outcomes. Learners were provided with two learning content types (video, PDF text) for every learning topic. It was up to the learner to select learning content types of her/his choice during the learning process. Providing two types of learning content to the learners was to know their preferences regarding learning content type. After completing a learning activity, the learners were encouraged to post any problem that they still have regarding a learning topic.

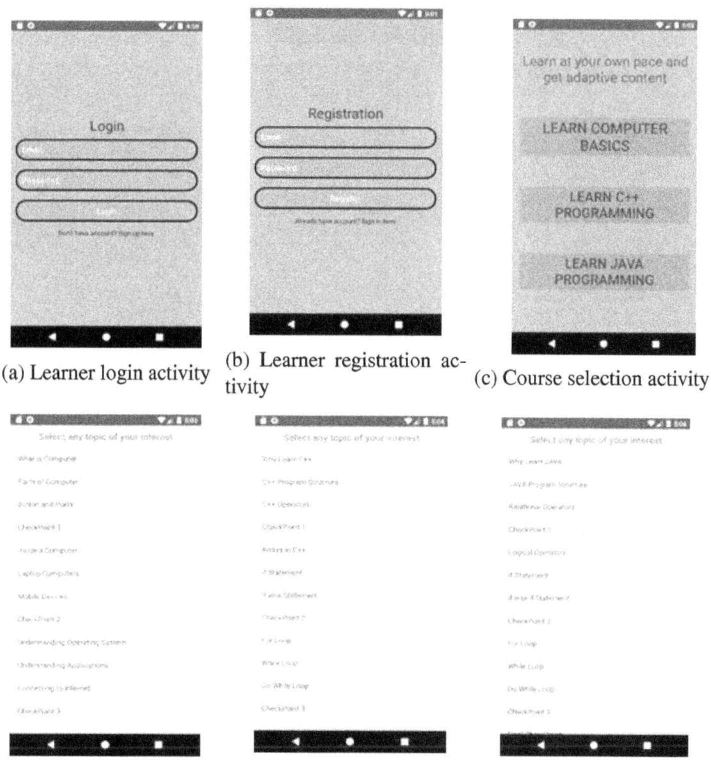

(a) Learner login activity (b) Learner registration activity (c) Course selection activity

(d) Computer Basics topics (e) C++ topics (f) JAVA topics

Figure 3.2: Learner Interaction with learnit app

3.1.3 Learnit Deployment Environment

Any learner having an Android smartphone with the minimum software development kit (SDK) 19 version could install and use it. The target SDK on which learnit was compiled and installed was Android Pie version 28. Table 3.1 lists the learnit app's

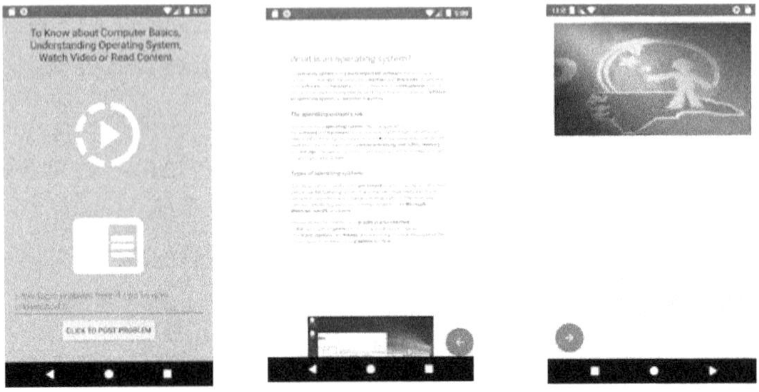

(a) Video or Text selection activity (b) Learner reading text content (c) Learner watching video content

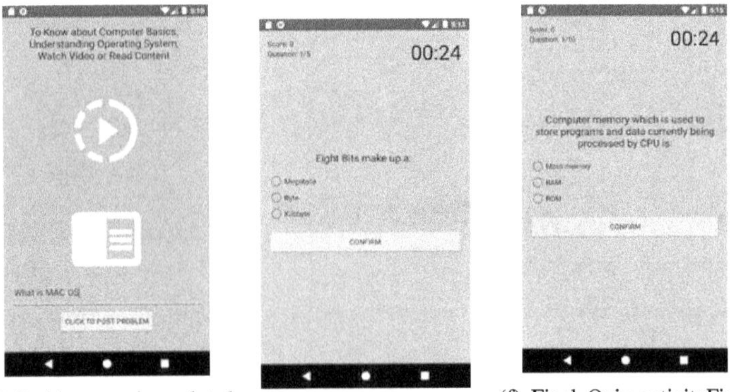

(d) Problem posting related to a particular topic. (e) Modules quiz activity (f) Final Quiz activityFinal Quiz activity

Figure 3.3: Learner learning, problem posting, and quiz activities

needs while the learner is interacting with it.

On the backend, the learnit was supported by Firebase, Google mobile, and web application development platform that provides several essential services to mobile apps. In our study, the key services that the Firebase platform provided us were learners' authentication (login and registration services), real-time database support, storage services (for learning content persistence), learners' activities analytics, learners' prediction (leaners behavior to an app) and cloud messaging. Firebase client SDK installed on Android Studio enabled us to interact with Firebase services directly with no need to establish middleware between the learnit app and the Firebase backend services. Database queries applied to Firebase real-time database were written at the learnit client side.

Table 3.1: Requirements of learnit app

R1	The learnit app should be able to automatically detect contextual features such as learner study time duration, the subject covered, the topic covered within a subject, learning time, repetition rate, learning content type i.e. video or text, problem posted online by a learner, quizzes attempts, quiz repetition rate, performance score. In some cases where it was impossible to automatically detect learning features, the features were collected manually from the learners.
R2	The learnit app should be able to stream or import learning content according to the preferences of learners (whether a learner is interested in reading text or watching videos).
R3	The learnit app should automatically log the learning features on the online Firebase cloud.
R4	The learnit app should also work when there is no Internet connection during the execution of learning activities. In this case, only the learning content already viewed or studied by a learner will be available.
R5	The learnit app should provide real-time database support to the learners. This means that any change to the learning content made by instruction would be instantly available to the learners.
R6	The learner should be able to view the graphical structure of learning activities and easily navigate to these learning activities.

While using Firebase services, the traditional backend is bypassed, putting database interaction responsibilities on the client app. The Firebase console provided administrative access regarding the learnit app. Figure 3.4 shows the difference between traditional backend services and Firebase services. Firebase's real-time database and storage services are somewhat different from the conventional backend database. They provide real-time updates to the client as the data changes in the database. Whenever the instructor uploads new learning material in our system, the update is instantly offered to learners at the client-side. The real-time database service from Firebase assured us that the learners are always using up-to-date learning material.

3.1.4 M-learners' features gathering

As mentioned earlier, the learners' features were extracted and gathered from learners while using the learnit app. The features were divided into three categories: behavioral features, context features, and final performance feature. The behavioral features manifest how the learner interacts with the learnit app, i.e., online discussion group participation, the number of times problem posted, the number of times problem solved, the number of times attempted quiz, and topic repletion rate. Similarly, context features

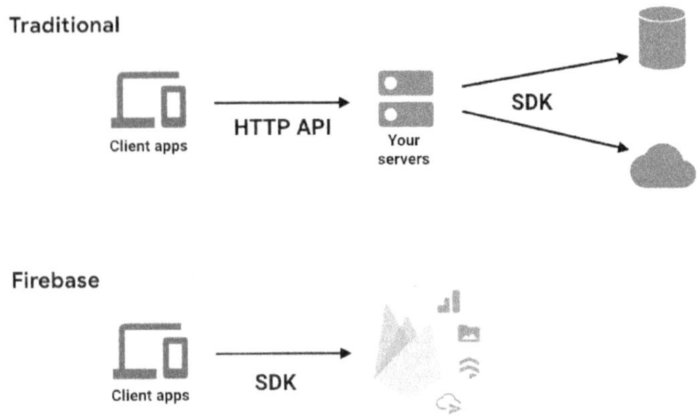

Figure 3.4: Traditional web services vs Firebase cloud approach

illustrate how learners used the learnit app in different contexts. The learning context comprises learner background knowledge, learning duration, learning time, places visited, learning content type (video, text), and modules 1, 2, 3 performances (depicts current learner knowledge about the topic). The final performance feature is about the learner's performance in the final checkpoint/quiz. The MCQs in the final checkpoint are from learning the content of module 1, module 2, and module 3. Table 3.2 shows the preprocessed features along with their category and description, whereas figure 3.5 shows the snapshot of the sample of features instances in CSV (Comma-Separated Values) format.

Table 3.2: Preprocessed features along with their category and description

Features Categories	Features	Description
Behavioral Features	ODGP	Online discussion group participation i.e. the number of times a learner has participated in the online discussion group (numeric)
	NPP	Number of times problem posted online related to study topics (numeric)
	NPS	Number of times solved the posted problem online by other learners during learning (numeric)

Table 3.2 continued from previous page

	NTAQ	Total number of quiz attempts in three modules (numeric)
	TRR	Learning topic average repetition rate in three modules (numeric)
Context Features	AST	Average study time per learning topic (numeric)
	NTRA	Number of times text resource accessed during the learning process (numeric)
	NVRA	Number of times video resource accessed during the learning process (numeric)
	MP1	Learner performance in module 1, numeric: (18 to 20 = very good), (15 to 18 = good), (12 to 15 = average), (9 to 12 = satisfactory), (0 to 9 = fail)
	MP2	Learner performance in module 2, numeric: (18 to 20 = very good), (15 to 18 = good), (12 to 15 = average), (9 to 12 = satisfactory), (0 to 9 = fail)
	MP3	Learner performance in module 3, numeric: (18 to 20 = very good), (15 to 18 = good), (12 to 15 = average), (9 to 12 = satisfactory), (0 to 9 = fail)
	APV	Academic Places Visited. Degree of academic places visited during the learning process i.e. library, classrooms, numeric: 0 to 1
	SPV	Social Places Visited. Degree of social places visited during the learning process i.e. playgrounds, hostels, numeric: 0 to 1

Table 3.2 continued from previous page

Final Performance Feature	FG	Final Grades showing the final performance of a learner in three modules, Categorical: (18 to 20 = very good), (15 to 18 = good), (12 to 15 = average), (9 to 12 = satisfactory), (0 to 9 = fail)

ODGP	NPP	NPS	NTAQ	TRR	AST	NTRA	NVRA	MP1	MP2	MP3	APV	SPV	FG
5	7	4	3	3	4	2	1	17	16	16	0.5	0.5	17
4	2	3	2	2	4	1	1	16	15	14	0.6	0.4	15
3	2	2	1	2	3	1	1	14	13	15	0.7	0.3	14
2	2	3	2	3	4	2	1	13	12	14	0.6	0.4	15
3	3	2	2	4	5	2	2	15	16	17	0.7	0.3	16
1	2	1	1	2	3	1	1	12	14	15	0.5	0.5	15
2	2	1	2	3	3	1	1	13	13	14	0.8	0.2	12
1	1	0	2	1	4	1	0	12	11	12	1	0	13
2	2	1	1	2	3	1	1	15	14	16	0.6	0.4	15
2	0	1	2	2	4	1	1	16	16	17	0.9	0.1	18
1	1	1	2	2	3	1	1	14	13	15	0.8	0.2	16
2	1	0	1	2	3	2	0	15	15	14	0.9	0.1	15
1	1	1	1	1	2	1	0	12	13	14	0.8	0.2	15
2	1	0	2	3	4	1	2	13	15	15	0.7	0.3	16
0	0	0	3	2	3	1	1	14	14	16	0.8	0.2	17
3	2	1	1	2	3	1	1	15	14	15	0.7	0.3	16
1	2	2	0	1	4	1	0	13	14	15	0.5	0.5	15
3	4	1	2	2	4	1	1	14	14	15	0.7	0.3	17
2	3	1	2	2	4	1	1	15	15	16	0.9	0.1	16
2	0	1	1	3	4	1	2	16	15	17	0.9	0.1	17
2	0	2	3	1	5	1	0	15	16	17	0.9	0.1	16
0	1	1	1	1	3	1	0	16	17	17	0.7	0.3	17
1	2	2	1	2	3	1	1	17	17	15	0.8	0.2	16
1	4	0	3	2	3	1	1	14	16	17	0.8	0.2	17
2	4	0	3	3	4	1	2	16	17	18	0.8	0.2	18
1	3	1	3	1	3	1	0	16	15	14	0.7	0.3	17
2	3	2	3	1	4	0	1	15	16	17	0.8	0.2	17
2	0	1	0	1	2	1	0	16	15	17	0.8	0.2	16

Figure 3.5: Features instances in CSV format. These features instances for feed to the learner model for further processing.

The behavioral and context features are called independent features, whereas the final performance feature is called a dependent feature. The independent features will act as input to the learner model to reveal how they affect the learners' final performance. Our dataset is composed of 5 behavioral features, 6 context features, and 1 final performance feature. Module 1, 2, and 3 performance scores are derived from the average of three modules quizzes in computer basics, C++, and JAVA courses, respectively. Eight hundred seventy-five learners from the Institute of Computing, Department of Physics, and Institute of Numerical Science participated in using the learnit app. After data cleansing and security, 850 learner's data were finally selected to be included in the dataset and further used as an input to the learner model.

When the learner interacts with learnit, the learner's features are stored on Firebase real-time database in JSON (JavaScript Object Notation) format. The JSON format is commonly used to store and transport data from online clouds. The learner remains synchronized with the real-time database throughout the learning process. Firebase provides a reliable, secure, and fast mechanism to synchronize data with users with minimal effort both on users and developers. Firebase real-time database cloud also provides data persistence by storing data locally on learners' mobile phones, thereby enabling them to access and use learning content even if the mobile phone device is offline. When the learner connection is re-established with the Firebase database, the up-to-date learning contents are automatically merged and synchronized with remote learner data. JSON data type is made up of key/value pairs. The purpose of the key is to identify the object within the database uniquely, and the value represents the actual value of the object. Multiple JSON objects are stored in the form of the parent/child tree. Figure 3.6 shows the pictorial representation of how learner information is stored in JSON format.

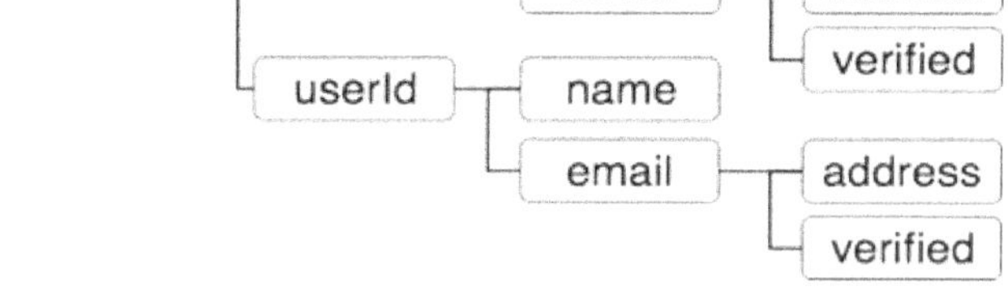

Figure 3.6: Learners' features stored initially in JSON format at Firebase real-time database.

After using learnit for two months, the learners' features recorded on Firebase cloud were downloaded and converted into CSV (Comma-separated Values) format. The features were further preprocessed and converted into a form that the learner model can accept as an input. The preprocessed features represent the real values of learners' features in the form of a set represented as <>. When processed by the learner model, these features will divulge which features contribute most to the learning process and how learners can be categorized into different groups based on their performance. The

Table 3.3: One-hot encoding applied to learners final grades

Final Grades	A	B	C	D	F
A	1	0	0	0	0
B	0	1	0	0	0
C	0	0	1	0	0
D	0	0	0	1	0
F	0	0	0	0	1

learner model also reveals the effect of independent features on the dependent feature, i.e., learners' final performance. The features preprocessing steps include features encoding features scaling and splitting features into training and testing sets.

3.1.5 Features Encoding

We are confronted with a learning classification task as the learner model will classify learners into different groups based on their performance. We trained the learner model to classify learners into 5 categorical groups, namely A, B, C, D, and F, representing their final performance grades. Before feeding learners' features to the learner model, it is necessary to encode them into an appropriate form acceptable to the learner model. Looking at our dataset, we noticed that luckily, we have one categorical feature, i.e., final grades, that need to be encoded. We used a technique called one-hot encoding to encode learners' final grades into numeric form. The one-hot encoding technique is used because DL and ML-based models do a better job in prediction and classification tasks with this technique. The one-hot encoding technique also allows the DL and ML-based models to break the ordinal association among categorical features. Due to high numerical value, one categorical feature does not get preference and larger weight over other categorical features. Table 3.3 shows the one-hot encoding technique applied to learners' final grades.

3.1.6 Features Scaling

All features are converted into one unit to ease the learner model processing and classification task (scaled or standardized). Features scaling is a necessary step as it would compel the learner model not to gives higher weightage to one independent feature over other independent features. Features scaling also facilitate DL and ML-based models in parallel and high computation tasks. After performing features scaling, all independent features are normalized and are treated equally by the learner model. The values of features are divided by the highest value for each feature to perform features scaling, as shown in the below equation 3.1.

$$X_{\text{norm(1-13)}} = \frac{X_1}{MAX(X_1)}, \frac{X_2}{MAX(X_2)}, \ldots\ldots, \frac{X_{13}}{MAX(X_{13})} \tag{3.1}$$

3.1.6.1 Splitting Features into Training and Testing sets

The dataset features instances were split into training and testing sets. Training set feature instances were used to train the learner model on the relationship between input independent features and output dependent features. The testing set features instances were used to test our learner model on new and unseen features instances. The testing set features instances that provide learner model accuracy and performance on new feature instances. The training and testing sets were made on a features dataset with 85:15 percent ratios.

In chapter 3, we discussed the framework, development, testing, and deployment of the learnit app, which is a part of a larger ILS. Among various purposes, one of the goals of the learnit app was to gather the learning features from M-learners initially. For in-depth analysis, the learning features were divided into behavioral and contextual features. After feature collection steps, we performed feature encoding, feature scaling steps. For training the M-learning model, the features were further divided into training and testing steps. In chapter 4, we will discuss the M-learning model's training using artificial neural network, M-learning model tuning, determining feature weights, and early intervention experiments.

CREATING LEARNER MODEL USING ANN

This chapter is further divided into 6 sections. Section 4.1 briefly discusses exploring and visualizing the correlation matrix among learning features. Section 4.2 describes how the M-learning model was trained using ANN. In section 4.3, the M-learning model's implementation process with ILS is discussed and how various M-learning models were used to determine their accuracy. Section 4.4 is about using the Random Forest (RF) algorithm to determine the feature weights and importance in increasing the performance of M-learners. In section 4.5, we discussed the early intervention experiment and why it was carried out on controlled and experimental group learners. Section 4.6 is about analyzing the results generated from the early intervention experiment.

After the features preprocessing step, we developed an m-learning model to classify learners according to their performance grades using ANN. The m-learning model is also capable of predicting new learners' performance given their past feature instances. The steps involved in building the m-learning model include:

- Formulating problem statement
- Gathering data
- Cleaning data
- Exploring and Visualizing data
- Training learner model
- Evaluating and testing the learning model

4.1 Exploring and Visualizing data

Exploring and visualizing data is essential as it enables us to understand what type of data we are working on and how meaningful it is to us. First, a correlation matrix was generated to reveal how the underlying features related to each other. The correlation matrix can be used to get a quick appreciation and understanding of how the underlying features interact. More importantly, the correlation matrix also tells us how independent features relate to learners' final grades. Looking at the correlation matrix in Figure 4.1, we observed that MP1, MP2, MP3 features have a strong correlation with the final grade feature whereas ODGP, NPP, and NTRA features have a good positive correlation with final grades. These figures tell us that out of 13 independent features, MP1, MP2, MP3, ODGP, NPP, and NTRA contribute more to increasing learners' final grades. The relationship of TRR with NTRA and NVRA. This relationship infers that learners consider revision (Topic Repetition Rate, TRR) as an essential factor in their learning process while accessing text or video learning material/resources.

Similarly, the correlation of the NPP feature with MP1, MP2, and MP3 implies that the more the learner has posted problems online, the better the result s/he achieves in MP1, MP2, and MP3. The correlation analysis also reveals that some features have a weak negative correlation with FG grade and other independent features. As we have a weak negative correlation, therefore we have ignored them and considered those features responsible for more considerable variance in the FG feature.

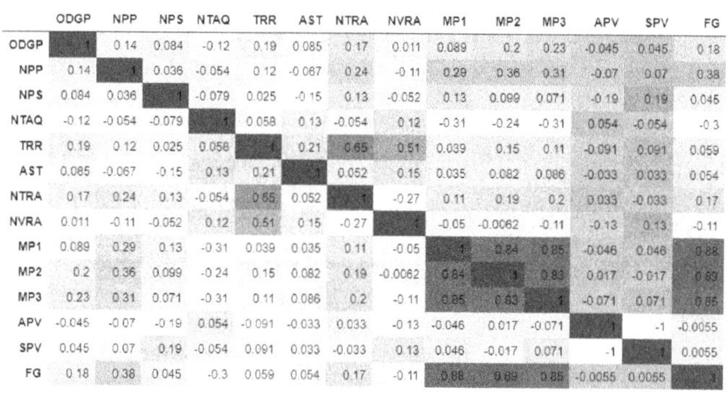

Figure 4.1: Correlation matrix of m-learning model features

4.2 ANN-based m-Learning Model (Modeling Learner using ANN)

To develop the m-learning model, ANN with two hidden layers was used as shown in Figure 4.2. For the ANN-based learning model, the most crucial element is historical feature instances. The more the feature instances are provided to the learner model, the better it will learn and generate accurate results. In recent times, ANN with hidden layers has gained tremendous success and considerable popularity with thousands of application domains such as fast and accurate information retrieval, user behavior analytics, handwriting recognition, stock exchange rates predictions, housing prices prediction, employee's retention prediction, student's dropout prediction, natural language processing, and adaptive websites, etc. For accurate prediction and classification tasks, ANN must be trained with historical features data and thoroughly sieved from overfitting and underfitting deficiencies. This way, the ANN-based model will generalize well and give reasonably good results on future and unseen data. In this study, the ANN-based learner model should accurately categorize learners into 5 (A, B, C, D, F) different groups based on their learning performance. This approach is called supervised deep learning classification and modeling. It is called supervised because input feature instances have known output labels or results, i.e., learners' final grades. It is called a classification problem as ANN categorizes learners into different groups based on their previous performance.

The ANN-based m-learning model shown in Figure 4.2 has 13 input neurons at the input layer. Each neuron at the input layer corresponds to a single input feature. Furthermore, there are two hidden layers with 11 neurons and one output layer having 5 neurons. The 5 neurons at the output layer are dedicated to showing the m-learning model result by classifying learners' performance into 5 grades. During the implementation, we used Python Sequential class to map input features to input layers neurons whereas the Python Dense class was used to initialize ANN edges with random weight values.

4.2.1 Training ANN-based m-learning model

Training the ANN-based m-learning model includes the following steps:

- Feeding the m-learning model with feature instances i.e., feed-forward technique or forward propagation where data flows in one direction, from input to output layer.

- Selecting the activation function for hidden layer neuron activation. In our experiment, we have selected the ReLU activation function.

- Using activation function at the output layer. As our problem is categorical, therefore, we have used the Softmax function.

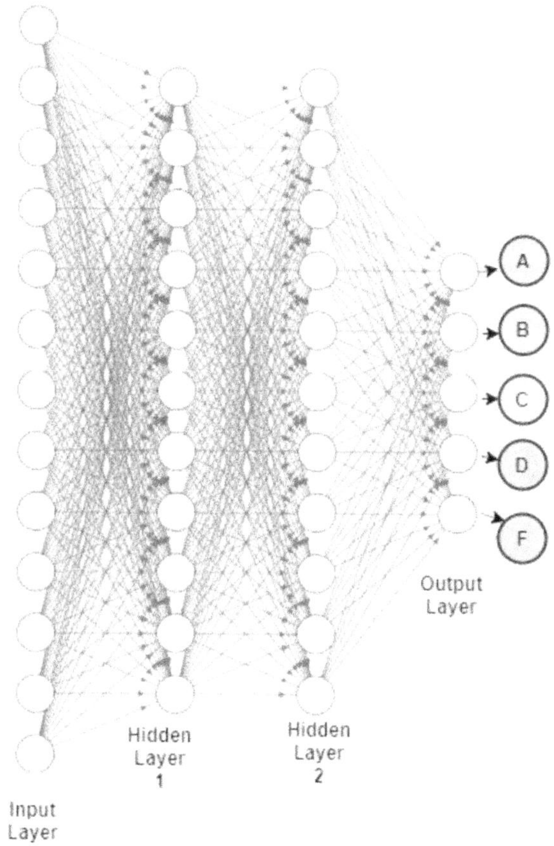

Figure 4.2: Artificial Neural Network based m-learning model with 2 hidden layers, 1 input, and 1 output layer.

- Back-propagation: To reduce the difference between the actual result and generated result i.e. to reduce the loss.

- Forward propagation is the core process during the model training and learning phase. It is the forward propagation and the backpropagation technique, where a model learns the weights and bias values for neural network synapses. As the name suggests, the input features are embedded into the input layer's network, flow through hidden layers, and generate output results at the output layer. The forward propagation is done in one direction from the input to the output layer. Our input features data are in the form of n * 13 where 13 are the number of input features, and n are feature instances/records. The default configuration of ANN accepts one feature instance simultaneously, but multiple feature instances can be entered into ANN by a technique called matrix multiplication. Using matrix multiplication, multiple feature instances can enter ANN at a time, speeding up the training and learning process as a result. To perform matrix multiplication, we defined two matrices, X and W1, where X represents input features data and W1 is the weight of the synapse between the input layer and hidden layer 1. The matrix X is N*M, where M is the input features and N are the feature instances.

Similarly, the W1 matrix represents the synapses' weights between the input layer and hidden layer 1, having 13*11 dimensions where 13 are the input neurons, and 11 illustrate the weights of the synapses attached to each input neuron (11 synapses per neuron). We used the Python Dense class to initially initialize weights on synapses. Later, synapses weights get updated according to the error value backpropagated from the output layer. Mathematically, the matrices X and W1 can be represented as:

$$X = \begin{bmatrix} X1_1,\ X2_1,\ X3_1, ...\ X13_1 \\ X1_2,\ X2_2,\ X3_2, ...\ X13_2 \\ X1_3,\ X2_3,\ X3_3, ...\ X13_3 \\ .... \\ X1_n,\ X2_n,\ X3_n, ...\ X13_n \end{bmatrix} W_1 = \begin{bmatrix} W1_1,\ W1_2,\ W1_3, ...\ W1_{11} \\ W2_1,\ W2_2,\ W2_3, ...\ W2_{11} \\ W3_1,\ W3_2,\ W3_3, ...\ W3_{11} \\ .... \\ W13_1,\ W13_2,\ W13_3, ...\ W13_{11} \end{bmatrix}$$

Matrix X, when multiplied by matrix W1, yields matrix a1 having n*11 dimensions as illustrated in equation 4.1. For simplicity the resultant matrix multiplication process in represented in equation 4.2.

$$a1 = \begin{bmatrix} \{X1_1 * W1_1 + ... + X13_1 * W13_1\}, \{X1_1 * W1_2 + ... + X13_1 * W13_2\}, ...\{X1_1 * W1_{11} + ... + X13_1 * W13_{11}\} \\ \{X1_2 * W1_1 + ... + X13_2 * W13_1\}, \{X1_2 * W1_2 + ... + X13_2 * W13_2\}, ...\{X1_2 * W1_{11} + ... + X13_2 * W13_{11}\} \\ \{X1_3 * W1_1 + ... + X13_3 * W13_1\}, \{X1_3 * W1_2 + ... + X13_3 * W13_2\}, ...\{X1_3 * W1_{11} + ... + X13_3 * W13_{11}\} \\ .... \\ \{X1_n * W1_1 + ... + X13_n * W13_1\}, \{X1_n * W1_2 + ... + X13_n * W13_2\}, ...\{X1_n * W1_{11} + ... + X13_n * W13_{11}\} \end{bmatrix} \tag{4.1}$$

$$a1 = XW_1 \tag{4.2}$$

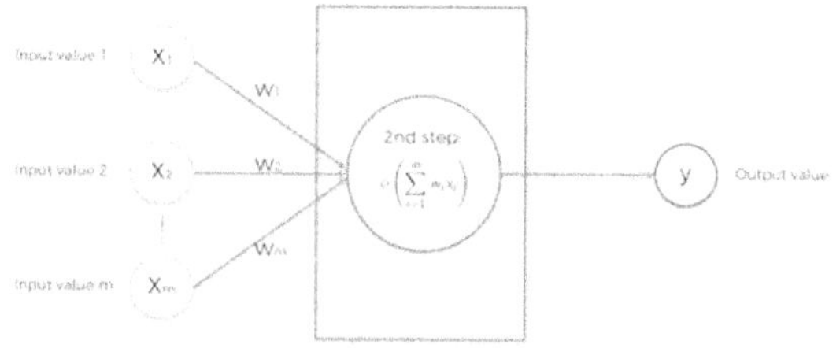

Figure 4.3: The simplified m-learning model learning process.

Two processes are performed during ANN forward propagation when input data is transferred from the input layer to hidden layer 1. First, at each neuron at hidden layer 1, the input feature value multiplies by synapses weights. This multiplication task is done at each neuron in hidden layer 1 for all features data and synapses connected to the corresponding neuron. The multiplication results are then added together, and the activation function is applied to the result. The following equation can summarize the processing at each neuron in hidden layer 1. θ represents the activation function performed after the sum of the product operation. Figure 4.4 illustrates a simplified and generic understanding of operation perform at each neuron in hidden layers.

$$\theta\left(\sum_{k=1}^{11} X_i W_i\right)$$

Secondly, the activation function is performed on the result obtained from the sum of the product operation. We used Rectified Linear Unit (ReLU) activation function at each hidden layer neuron because it is computationally less expensive, efficient and can easily propagate errors during the backpropagation process. ReLU activation function is represented in the equation 4.3.

$$z1 = f(a1) \tag{4.3}$$

where f(a1) is

$$f(a) = \begin{cases} 0 for a1 < 0 \\ a1 for a1 >= 0 \end{cases}$$

The ReLU activation function is applied independently to each entry in matrix a1, and a new matrix is generated called z1. The size of matrix z1 is the same as matrix a1. All the neurons at hidden layer 1 perform the ReLU activation function on their input matrix and generate modified matrices ready for feed-forwarding to hidden layer 2. The matrix data is propagated forward from hidden layer 1 to hidden layer 2 in the same way as it was propagated from the input layer to hidden layer 1. Similarly, the operation at hidden layer 2 neurons is the same as hidden layer 1 neurons operation. At hidden layer 2 neurons, the input values from hidden layer 1 are multiplied by corresponding synapses weights and then added together to perform the ReLU activation function. The z1 matrix stored at hidden layer 1 has an n*11 dimension, whereas the dimension of the W2 matrix is 11*11 where 11 represents neurons at hidden layer 1 and their corresponding weights. The multiplication of z1 matrix with W2 yields a2 matrix having the dimension of n*11 and can be represented by the following equation 4.4

$$a2 = z1W2 \tag{4.4}$$

After matrix multiplication operation, the ReLU activation function is performed on each entry in matrix a2 resulting in a new matrix called z2. Mathematically, the equation can be written as:

$$z2 = f(a2) \tag{4.5}$$

Similarly, using the forward propagation technique, the matrix z2 at hidden layer 2 is propagated to the output layer. The neurons at the output layer after performing softmax activation function (for categorical/multiple classification results) on input data, generate, $\hat{p}_j$, which is the ANN-based m-learning model predicted the value of the learner's final grade (FG). It is very likely that initially, there is a difference between the predicted result $\hat{p}_j$ and actual result 'a'. For minimizing the difference between the predicted result and the actual result, we used a technique called back-propagation, which is further explained in the following section.

4.2.2 Minimizing Model Cost/Loss using back-propagation

M-learning model cost or loss function tells us the difference between predicted and actual output result. Ideally, the predicted result should be equal or very near to the actual result but in a scenario where the difference is substantial, the back-propagation technique is used to reduce the difference between predicted and actual output results.

We expressed the learner model cost function in the following expression.

$$C = \sum \frac{1}{2}(a - \hat{p}_j)$$

In the above expression, 'a' is the learner's actual final grade whereas $\hat{p}_j$ is the predicted final grade result. For an accurate and reliable learner model, this difference must be as minimum as possible. As a result, the m-learning model will convey an accurate picture of learners' learning behavior and performance. For minimizing the learner model cost function, we had two choices: 1) changing input features data, 2) tweaking weights on ANN synapses. Input feature instances are static and will not change therefore the only option left with us is to adjust weights on ANN synapses. To determine suitable weights for synapses, we used a technique called Stochastic Gradient Descent (SGD) along with back-propagation. SGD technique is efficient and reduces time and computation costs during the back-propagation process. Furthermore, when we have high dimensional data and we are looking for suitable weights across millions of synapses, the SGD technique is more impressive in determining suitable synapses weights and therefore reducing the overall model cost. Looking at our ANN model architecture, we can observe that the weights that are responsible for generated cost/error are spread across three matrices i.e., W1, W2, W3. We calculated the derivative of cost C concerning W1, W2, W3 i.e., d(C)/d(W1), d(C)/d(W2), d(C)/D(W3), resulting in the same number of gradient values as the number of synapses weights across ANN. The derivative of C concerning W1, W2, W3 determines how much weight values are responsible for generating error/cost in the m-learning model.

The key in the training of the M-learning model is to find the appropriate values for weights across the model so that the model can build good approximator functions that can be applied to the new learner features for their performance prediction. In other words, we want to have a set of weights that minimize the error/loss/cost at the output layer. To demonstrate the back-propagation process in detail considers below the simple Figure 4.4 of the m-learning model.

If the weights w1 gets changed, the values across hidden layers 1 and 2, w1, w2, and the output result would also change. Here, the output result is functional dependent on w1, w2, w3 weight, and activation function in hidden layers 1 and 2. We can, mathematically, formulate the output as an extension of composite function:

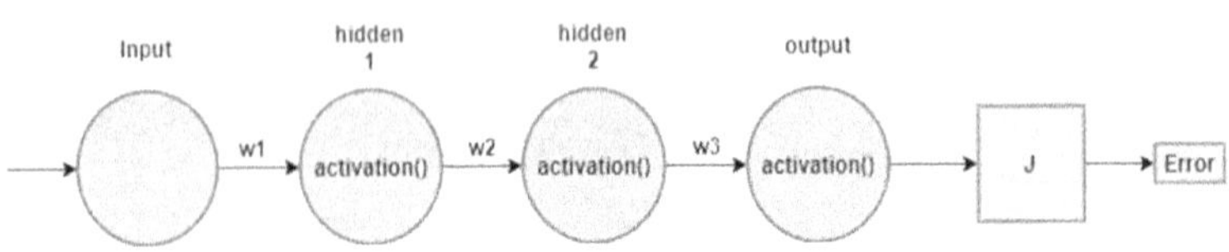

Figure 4.4: The simplified m-learning model learning process.

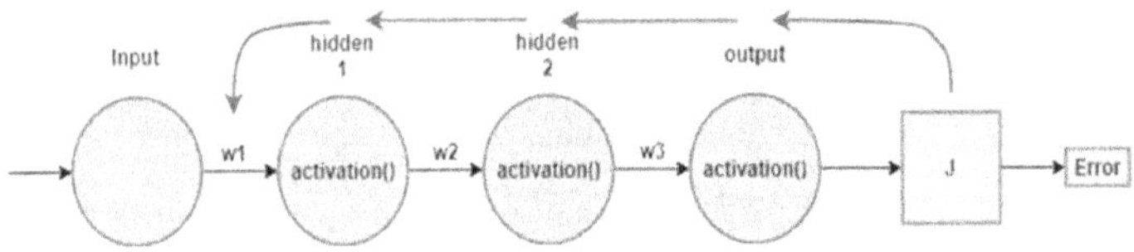

Figure 4.5: Backpropagation process where the generated error is backpropagated to weights w1.

$$output = act(w3 * act(w2 * act(w1 * input\_data))) \tag{4.6}$$

If we have an error on the output layer, then it means that the error is functionally dependent on the output function which in turn is dependent on the function of input, weights, and activation functions. To know how many arbitrary weights i.e., w1 is responsible for the generated error, we compute the derivative of the error w.r.t w1, and the result would be:

$$\frac{d(j)}{d(W_1)} = \frac{d(j)}{d(output)} \cdot \frac{d(output)}{d(hidden2)} \cdot \frac{d(hidden2)}{d(hidden1)} \cdot \frac{d(hidden1)}{d(W_1)} \tag{4.7}$$

In the backpropagation process, we are traversing from the output error to W1, taking iterative steps using the chain rule. We take the derivative of the error w.r.t the output, derivative of output w.r.t to hidden neurons, and finally derivative of hidden neurons w.r.t weights. This process is shown in Figure 4.5.

To get a comprehensive intuition of the backpropagation process, we are going to determine the derivative of error w.r.t the weights at the input layer denoted by $W^1{}_{ij}$ in Figure 4.6. The architecture of the learner model in Figure 4.6 is simplified for a better understanding of the backpropagation process. Mathematically, we are trying to obtain how much input layer weights are responsible for error as shown in equation 4.8.

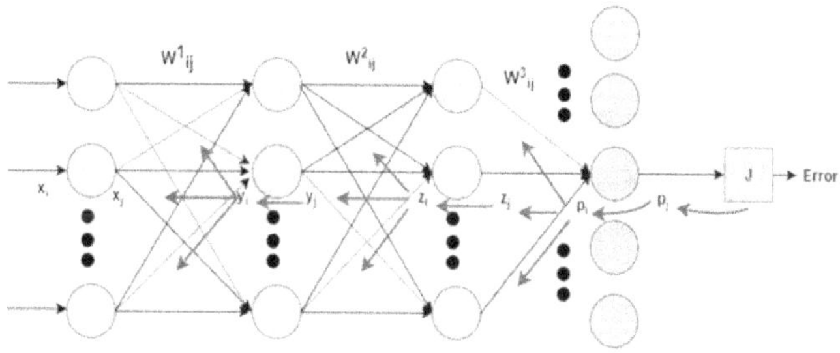

Figure 4.6: Simplified ANN-based m-learning model architecture.

$$\frac{d(error)}{d({W^1}_{\mathrm{ij}})} = \frac{d(j)}{d({W^1}_{\mathrm{ij}})} \tag{4.8}$$

Looking at the Figure 4.6, we also know that

$$\frac{d(error)}{d({W^1}_{\mathrm{ij}})} = \frac{d(j)}{d({W^1}_{\mathrm{ij}})} = \frac{d(j)d(p_{\mathrm{j}})}{d(p_{\mathrm{j}})d({W^1}_{\mathrm{ij}})} \tag{4.9}$$

This implies that the derivative of output error w.r.t to input layer weights is equal to the derivative of error w.r.t predicted result multiplied by the derivative of predicted result w.r.t input layer weights. The term d(j)/d(p_j) tells us the derivative of error w.r.t of output prediction which can be written as:

$$(P_{\mathrm{j}} - a) \ where \ \ a \ is \ the \ actual \ output \ result$$

Therefore, the equation becomes.

$$\frac{d(j)}{d({W^1}_{\mathrm{ij}})} = (P_{\mathrm{j}} - a)\frac{d(p_{\mathrm{j}})}{d({W^1}_{\mathrm{ij}})} \tag{4.10}$$

Going one layer further back and applying the chain rule we can write $\frac{d(p_{\mathrm{j}})}{{W^1}_{\mathrm{ij}}}$ as:

$$\frac{d(p_{\mathrm{j}})}{d({W^1}_{\mathrm{ij}})} = \frac{d(p_{\mathrm{j}})}{d(p_{\mathrm{j}})}\frac{d(p_{\mathrm{j}})}{d({W^1}_{\mathrm{ij}})} \tag{4.11}$$

From the above equation 4.11, we can deduce that the derivative of predicted result w.r.t input weights is equal to the derivative of output w.r.t to derivate of input (pi) at the output layer multiplied by the derivative of input value w.r.t input weights. Similarly, applying the chain rule $\frac{d(p_{\mathrm{j}})}{d(p_{\mathrm{j}})}$ can be written as:

$$\frac{d(p_j)}{d(p_j)} = p_j(1 - p_j)$$

In the above expression, the term at the right-hand side corresponds to the derivative of the ReLU function which is equal to the output result multiplied by one minus the output. Pj denotes the output of the activation function at the output layer. Now the equation 4.11 can be written as:

$$\frac{d(p_j)}{d({W^1}_{ij})} = p_j(1 - p_j)\frac{d(p_i)}{d({W^1}_{ij})}$$

$$\frac{d(j)}{d({W^1}_{ij})} = (p_j - 1) * p_j(1 - p_j)\frac{d(p_i)}{d({W^1}_{ij})} \tag{4.12}$$

Moving backward to hidden layer 2, we can notice that the red lines spread out in multiple directions. Since multiple weights values affect the value of pi, we must take all of them into derivative which can be represented by sigma notation:

$$\frac{d(p_j)}{d({W^1}_{ij})} = \sum_j \frac{d(p_i)}{d(z_j)}\frac{d(z_j)}{d({W^1}_{ij})}$$

Looking at Figure 4.6, we can observe that pi is the summation of hidden layer 2 weights multiplied by zj (activation performed on the sum of products). Taking the derivative of pi with zj would give us connecting weights between hidden layer 2 and output layer and can be expressed as:

$$\frac{d(p_i)}{d(z_j)} = {W^3}_{ij}$$

We can express the rate of change of p_i and z_j w.r.t ${W^1}_{ij}$ as:

$$\frac{d(p_j)}{d({W^1}_{ij})} = {W^3}_{ij}\frac{d(z_j)}{d({W^1}_{ij})} \tag{4.13}$$

Since we cannot quite put the derivative of z_j w.r.t ${W^i}_{ij}$ into a numerical form, we continue to use chain rule and following the fourth red line back in Figure 4.6 to determine that:

$$\frac{d(z_j)}{d({W^1}_{ij})} = \frac{d(z_j)}{d(z_i)}\frac{d(z_i)}{d({W^1}_{ij})} = z_j(1 - z_j)\frac{d(z_i)}{d({W^1}_{ij})} \tag{4.14}$$

Since z_j is the result of applying the activation function to z_i, therefore, we use the same logic as in the previous layer and apply the ReLU derivative. The derivative of z_i

w.r.t $W^1{}_{ij}$ established by the fifth line back is based on the same idea of spreading out and taking the summation of all contributions:

$$\frac{d(z_i)}{d(W^1{}_{ij})} = \sum_j \frac{d(z_i)}{d(y_j)} \frac{d(y_j)}{d(W^1{}_{ij})} = \sum_j W^2{}_{ij} \frac{d(y_j)}{d(W^1{}_{ij})}$$

We move on to performing more reduction and find the derivate of y_j w.r.t $W^1{}_{ij}$.

$$\frac{d(y_j)}{d(W^1{}_{ij})} = \frac{d(y_j)}{d(y_i)} \frac{d(y_i)}{d(W^1{}_{ij})} = y_j(1 - y_j) \frac{d(y_i)}{d(W^1{}_{ij})} \tag{4.15}$$

Traversing further back in our network, we notice that at red line number six in Figure 4.6, we still must deal with another activation function. The last line in the backpropagation process traverses from the input at y_i to x_j which represent the weights $W^1{}_{ij}$ between input and hidden layer 1. In fact, $W^1{}_{ij}$ are the weights which we were attempting to backprop to. At this point, we can determine the change in y_i w.r.t $W^1{}_{ij}$ which can be expressed by the following equation.

$$\frac{d(y_i)}{d(W^1{}_{ij})} = \sum_j x_j$$

Notice here that the coefficient of $W^1{}_{ij}$ is x_j. Combining all the individual expression gives us the final expression. Here the final expression tells us that how many weights to adjust at each layer to reduce to generated error.

$$\frac{d(j)}{d(W^1{}_{ij})} = (p_j - a) * (p_j(1 - (p_j) * \sum_j \left(W^3{}_{ij} * (z_j(1 - (z_j) * \sum_j \left(W^2{}_{ij} * (y_j(1 - (y_j) * \sum_j x_j \right) \right) \tag{4.16}$$

Initially, during the training process, the m-learning model may not find the right relationship between independents features and learners' final grades. Initially, the m-learning model may not be able to assign appropriate weights to the m-learning model synapses. The backpropagation technique gives suitable weights to synapses that provide an optimal estimation of a function that correctly models the training data through the iterative process. The weights assigned to synapses during the backpropagation process minimize the output error, ensuring that the difference between actual and predicted output is minimal. Equation 15 tells us about the amount of error to be back-propagated through the entire model. At each layer, during the backpropagation process, the weights of the updated synapses depend on the amount of error

Table 4.1: Learners' performance scores classified into five-level grades

Final Grades Classification	Very Good	Good	Average	Satisfactory	Fail
Earned Points	18-20a	15-17	12-14	9-11	0-8
Grades	A	B	C	D	F
a, the number is inclusive in the range					

back-propagated and how much stochastic gradient descent (SGD) technique generates new values for synapses weights. SGD updates the weights of synapses after every new instance inserts into the model. Therefore, it is more flexible, has higher fluctuation, and can find global minimum values for weights. SGD technique makes sure that the m-learning model never falls into a local minimum trap, which has a high negative effect on the model's accuracy and performance.

4.3 Implementation of the m-learning model using Artificial Neural Network (ANN)

We developed and implemented the m-learning model using ANN. Firstly, we initialized the weights of synapses with numbers very close to zero with the Keras Dense class's help. Secondly, the Keras Sequential class was used to propagate dataset features observations from the ANN input layer to the output layer. During forward propagation, the activation function at every neuron in ANN is restricted by the synapses' weights. The m-learning model's output is learners' performance grades, i.e., A, B, C, D, and F. Table 4.1 shows learners' performance scores classified into five-level grades.

The goal of developing the m-learning model and performing learners' multi-class classification was to classify learners into the right grades when their learning records/observations are feed into the m-learning model. Once the m-learning model is trained on well enough observations, it would classify new learners into the right grades in the future. During m-learning model development, the Keras library and API provided us fully configurable functions that can be plugged together with little effort and restrictions. Keras library provided ANN initialization schemes, cost function optimizers, regularization schemes, and activation function modules that we combined and created an m-learning model with high accuracy and performance. We assumed that features such as module performance 1 (MP1), module performance 2 (MP2), and module performance 3 (MP3) has the highest impact on the learners' final grades; therefore, we performed multi-class classification on four learning models as discussed below briefly:

- m-learning model 1: This model contains all the features from the dataset except final grades.
- m-learning model 2: This model is similar to m-learning model 1 but without the MP3 feature.
- m-learning model 3: This model is similar to m-learning model 2 but without the MP2 feature.
- m-learning model 4: This model is similar to m-learning model 3 but without the MP1 feature.

Before creating m-learning model 1, we performed features preprocessing, e.g., dataset cleaning, assigning grade categories (A, B, C, D, F) to the final performance, and splitting the dataset into training and testing sets. We also performed features scaling, i.e., normalization, so that one independent feature, due to higher numerical value, does not dominate other numerical features. Features scaling also speeds up neural networks' parallel computation process. Lastly, in the features preprocessing step, we encode categorical features into a suitable form. In our features' dataset, luckily, there is one categorical feature, i.e., final grades, that need to be encoded. To allow artificial neural networks to do better jobs with learners' classification and performance prediction, we used the one-hot encoding scheme to encode learners' final grades. One-hot encoding scheme breaks the ordinal relationship among categorical features, and therefore, due to higher numerical values, one categorical feature does not get preference over other numerical features. We used all independent features to predict final grades in m-learning model 1. The m-learning model 1 comprises one input layer with 13 input neurons (for accepting input features), two hidden layers having 11 neurons each, and one output layer for generating final grades.

The five output neurons in m-learning model 1 correspond to the five-level grading scheme. At hidden layers 1 and 2 neurons, ReLU (Rectifier Linear unit) activation function was used, whereas the Softmax activation function was used in the output layer to predict final grades. The hypermeters used in the creation of m-learning model

```
Layer (type)                 Output Shape              Param #
=================================================================
dense_13 (Dense)             (None, 11)                154
_________________________________________________________________
dense_14 (Dense)             (None, 11)                132
_________________________________________________________________
dense_15 (Dense)             (None, 5)                 60
=================================================================
Total params: 346
Trainable params: 346
Non-trainable params: 0
_________________________________________________________________
130/130 [==============================] - 0s 234us/step
Accuracy of the Model 90.77%
```

Figure 4.7: M-learning model 1 layers, output shape, parameters, and accuracy.

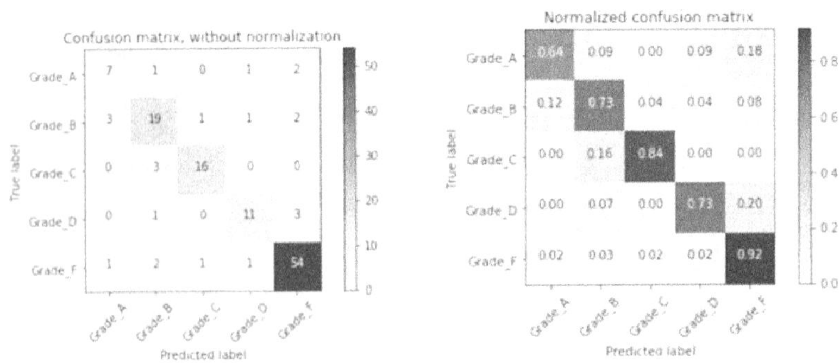

(a) Without normalization

(b) With normalization

Figure 4.8: Confusion matrices of m-learning model 1, without normalization and normalized confusion matrix.

1 were: epochs = 200, batch size = 5, optimizer = Adam, metrics = accuracy, loss = categorical_crossentropy, kernel_regularizer = regularizer.L2. With Keras and Tensorflow libraries' help, we compiled m-learning model 1 and achieved an accuracy of 90.77 percent, as shown in Figure 4.7. To better analyze and visualize the results of m-learning model 1, we constructed a confusion matrix on actual grade scores and predicted grade scores shown in Figure 4.8. (without normalization and normalized confusion matrices). The confusion matrix for m-learning model 1 revealed that 107 out of 130 grades were predicted correctly, whereas the remaining 23 grades were mispredicted.

Like m-learning model 1, m-learning models 2, 3, and 4 were constructed and after compilation, the models achieved an accuracy of 87.69, 83.85, and 80.00 percent respectively as depicted in Figure 4.9, Figure 4.11, and 4.15. The confusion matrices

```
Layer (type)                 Output Shape              Param #
=================================================================
dense_22 (Dense)             (None, 11)                143
_________________________________________________________________
dense_23 (Dense)             (None, 11)                132
_________________________________________________________________
dense_24 (Dense)             (None, 5)                 60
=================================================================
Total params: 335
Trainable params: 335
Non-trainable params: 0
_________________________________________________________________
130/130 [==============================] - 0s 227us/step
Accuracy of the Model 87.69%
```

Figure 4.9: M-learning model 2 layers, output shape, parameters, and accuracy.

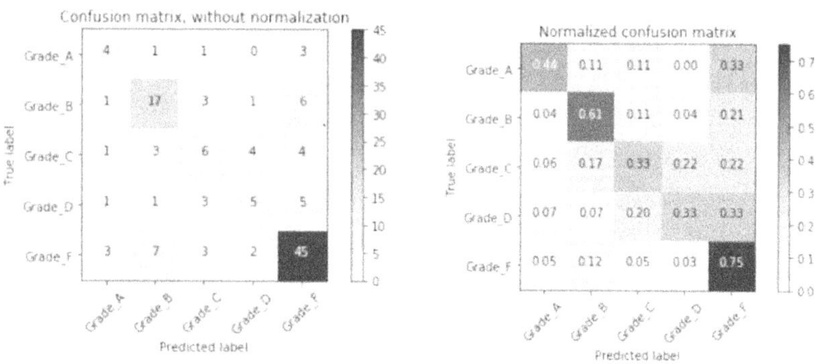

(a) Without normalization

(b) With normalization

Figure 4.10: Confusion matrices of m-learning model 2, without normalization and normalized confusion matrix

```
Layer (type)                 Output Shape              Param #
=================================================================
dense_25 (Dense)             (None, 11)                132
_________________________________________________________________
dense_26 (Dense)             (None, 11)                132
_________________________________________________________________
dense_27 (Dense)             (None, 5)                 60
=================================================================
Total params: 324
Trainable params: 324
Non-trainable params: 0
_________________________________________________________________
130/130 [==============================] - 0s 238us/step
Accuracy of the Model 83.85%
```

Figure 4.11: M-learning model 3 layers, output shape, parameters, and accuracy.

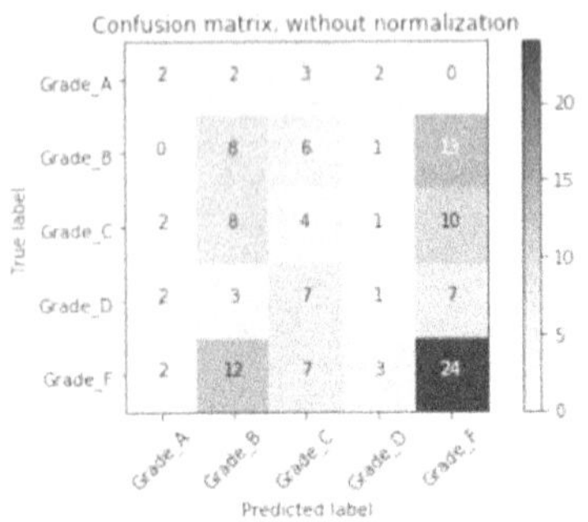

(a) Without normalization

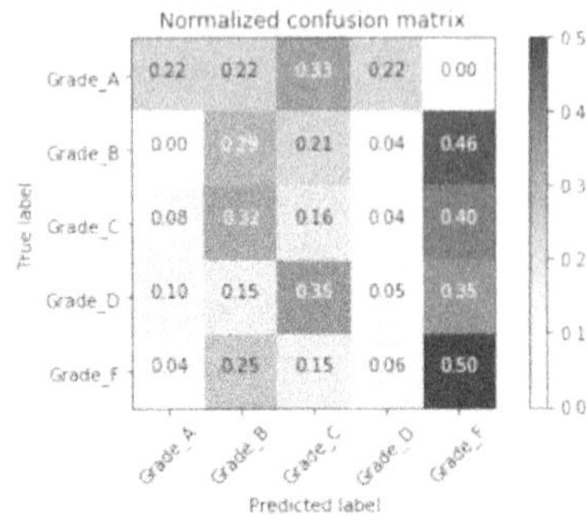

(b) With normalization

Figure 4.12: Confusion matrices of m-learning model 3, without normalization and normalized confusion matrix

```
Layer (type)                 Output Shape              Param #
=================================================================
dense_28 (Dense)             (None, 11)                121
_________________________________________________________________
dense_29 (Dense)             (None, 11)                132
_________________________________________________________________
dense_30 (Dense)             (None, 5)                 60
=================================================================
Total params: 313
Trainable params: 313
Non-trainable params: 0
_________________________________________________________________
130/130 [==============================] - 0s 238us/step
Accuracy of the Model 80.00%
```

Figure 4.13: M-learning model 4 layers, output shape, parameters, and accuracy.

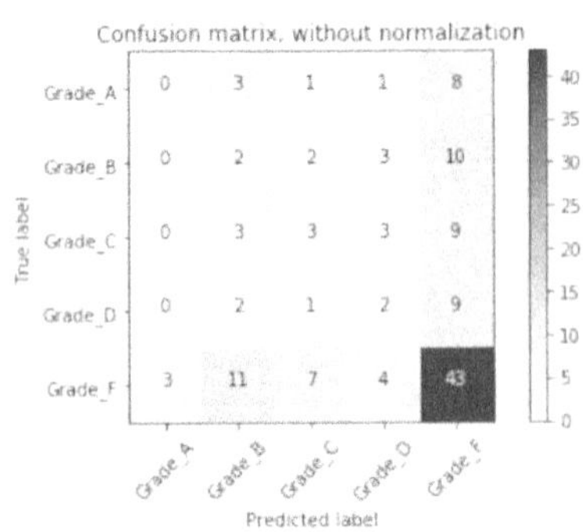

(a) Without normalization

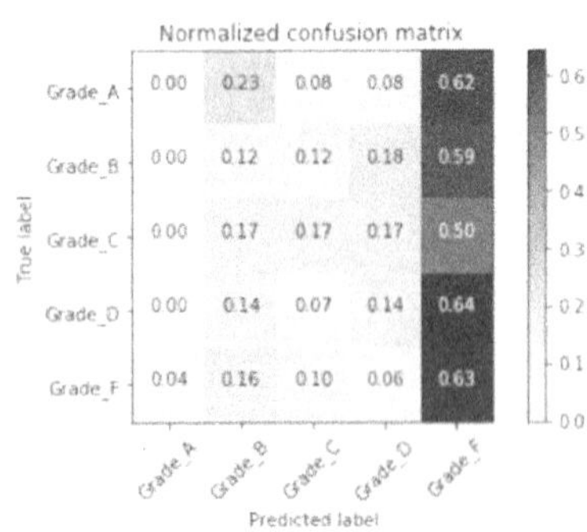

(b) With normalization

Figure 4.14: Confusion matrices of m-learning model 4, without normalization and normalized confusion matrix

```
Layer (type)                 Output Shape              Param #
=================================================================
dense_28 (Dense)             (None, 11)                121
_________________________________________________________________
dense_29 (Dense)             (None, 11)                132
_________________________________________________________________
dense_30 (Dense)             (None, 5)                 60
=================================================================
Total params: 313
Trainable params: 313
Non-trainable params: 0
_________________________________________________________________
130/130 [==============================] - 0s 238us/step
Accuracy of the Model 80.00%
```

Figure 4.15: M-learning model 4 layers, output shape, parameters, and accuracy.

(without normalization and normalized) for m-learning models 2, 3, and 4 are shown in Figure 4.10, Figure 4.12, and Figure 4.14. The diagonal elements in confusion matrices are those elements that are predicted accurately, whereas the remaining elements are predicted inaccurately. Those predicted accurately are called True Positives, whereas those predicted inaccurately are called False Positives.

From confusion matrices in Figure 4.10, Figure 4.12, and Figure 4.14, we can observe that the accuracy of models decreases as we remove the modules' performance i.e., the features MP1, MP2, and MP3 from independent features. This concludes that module performance scores are an important factor in predicting the final grades of learners. Moreover, to know whether our m-learning models suffer from overfitting, we plot training accuracy, testing accuracy, training loss, and testing loss over the number of epochs passed.

Figure 4.16, Figure 4.17, Figure 4.18, and Figure 4.19 shows the model accuracy and model loss for m-learning models 1, 2, 3, and 4. Figure 4.16, initially, during early epochs, the model accuracy of the training and testing dataset somewhat matched up. Still, later on, during the last 50 epochs, the model accuracy of the training and testing dataset somewhat deviates. Still, overall, it does not seem that overfitting is a huge problem for m-learning model 1, and the accuracy of the model is acceptable. The model loss of m-learning model 1 decreases as the number of epochs passes away.

This concludes that the m-learning model 1 successfully minimizes cost/loss function over time and is gradually learning the exact relationship between independent features and final grades. The results of the model accuracy and the model loss for

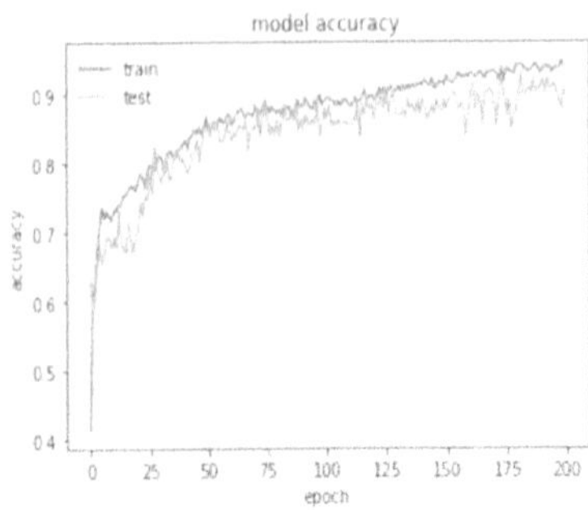

(a) Model 1 Accuracy

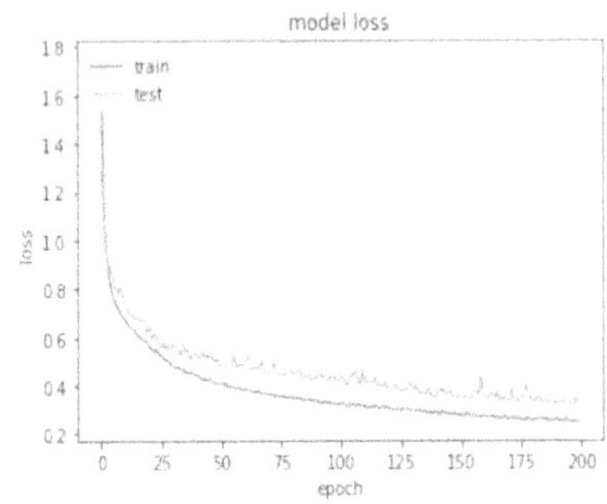

(b) Model 1 Loss

Figure 4.16: Model accuracy and model loss for m-learning model 1

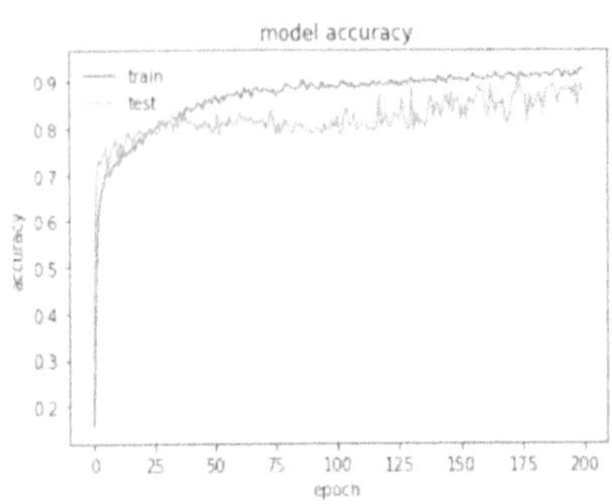

(a) Model 2 Accuracy

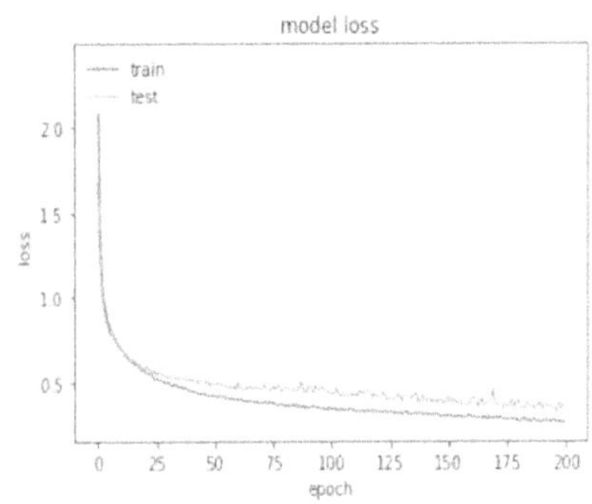

(b) Model 2 Loss

Figure 4.17: Model accuracy and model loss for m-learning model 2

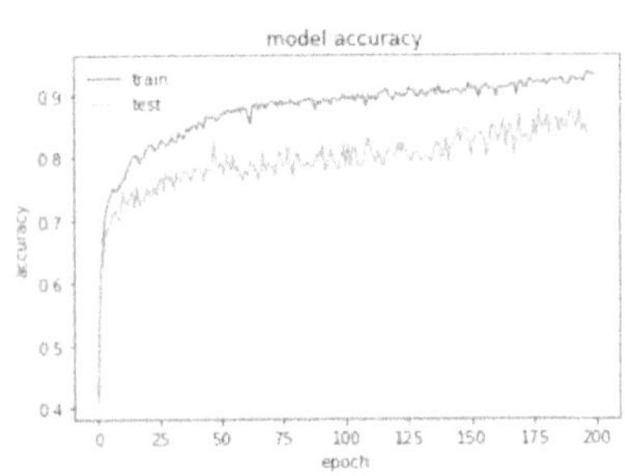

(a) Model 3 Accuracy

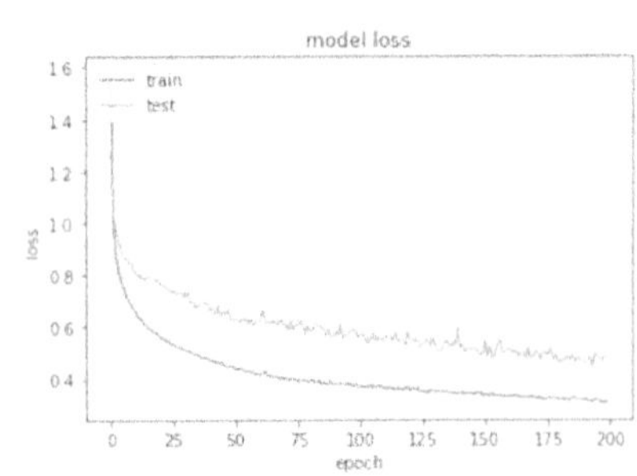

(b) Model 2 Loss

Figure 4.18: Model accuracy and model loss for m-learning model 3

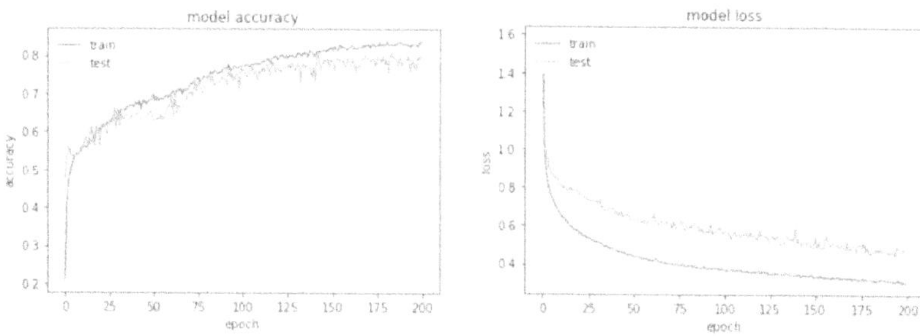

(a) Model 4 Accuracy (b) Model 4 Loss

Figure 4.19: Model accuracy and model loss for m-learning model 4

m-learning model 2 and m-learning model 3 are somewhat similar to m-learning model 1 as shown in Figure 4.17 and Figure 4.18 but we can observe that the case for model accuracy and model loss for m-learning model 4 is different. Figure 4.19 shows that as more and more epochs are completed, the difference between the model accuracy and model loss of training set and testing set increases.

This is because now, due to the absence of module performance features (MP1, MP2, MP3), the m-learning model 4 cannot predict the final grades and the model loss increases, and the model accuracy decreases on the testing set. The results in Figure 4.19. also divulged that module performance features are important in predicting the overall final grades of learners.

4.3.1 Multi-Class benchmark machine learning algorithms prediction accuracy in comparison with ANN

Before fitting all four learning models to different machine learning algorithms, some preprocessing was performed. The derived (from final performance feature) nominal feature of learner grades (A, B, C, D, F) was transformed into a one-hot-encoding scheme, and all other features were standardized to zero mean and one standard deviation. The parameters adopted for artificial neural networks were 200 epochs and a batch size of 5, whereas a total sample of 200 trees was selected for the random forest ensemble method. For support vector machines, we selected Sequential Minimal Optimization (SMO) algorithm. To estimate the predictive accuracy of each algorithm, the 5-fold cross-validation resampling technique was used. Under such a scheme, the data is randomly divided into 5 subsets of equal size. Subsequently, out of 5 subsets, one set is used as validation data for testing the model while the remaining 4 subsets are used as a training set. Using the 5-fold cross-validation technique, each subset of data is considered a test set once and at the end, the evaluated test sets contain the whole dataset. Models having a value of 5 generally result in lower bias and modest variance. Machine learning algorithms use the 5 variations of the dataset to create accurate predictions. From Table 4.2 it can be noted that ANN and Random Forest algorithms have superior prediction accuracy for all four learning models while linear discriminant analysis exhibited inferior accuracy.

Table 4.2: Accuracy results of machine learning algorithms and ANN on five-level grades classification.

Models/ Algorithms	Linear Discriminant Analysis (LDA)	Multi-class Logistic Regression (Softmax Regression)	Support Vector Machine (SVM)	Decision Trees (DT)	Random Forest (RF)	Artificial Neural Network (ANN)
Learning Model 1	85.56	88.67	87.56	88.56	90.11	90.77
Learning Model 2	78.55	79.56	79.44	76.59	86.89	87.69
Learning Model 3	67.78	70.45	74.45	67.43	83.67	83.85
Learning Model 4	54.12	55.98	59.64	56.89	80.23	80.00

4.4 Features weights and Importance

We used the Random Forest ensemble method to determine each feature's importance and weight in predicting the final performance grades. Random forest is based on an adaptive learning methodology that combines several models to generate better results than the individual model result. Random forest gives accuracy, robustness, easy interpretability, and control over underfitting and overfitting problems. Moreover, the random forest method helps the model in decreasing their variance, biasness, and increasing their prediction accuracy. The random forest method averages the results of

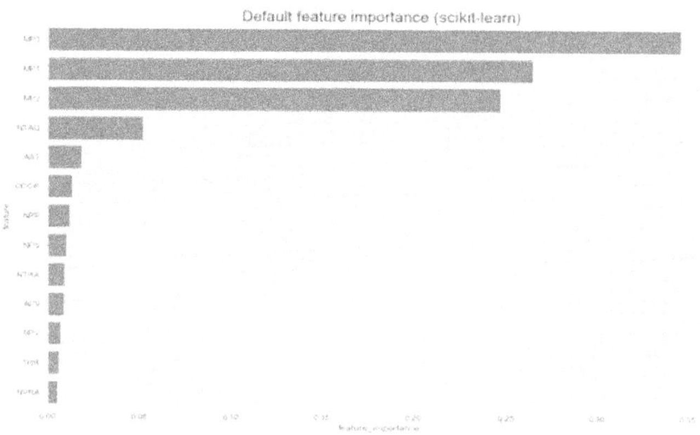

Figure 4.20: Features importance calculated with the help of scikit-learn feature_importances_ function.

the number of decision trees and predicts more accurate results by voting in case of classification problems and averaging in case of regression problems.

Feature weights and importance give satisfactory evidence of what features contribute most in predicting learners' final grades. The features weights were integrated with the m-learning model to provide adaptive content and tailored navigation to learners during their interaction with the learnit app. Integrating feature weights will make learnit a more comprehensive and complete application that will have early knowledge of learners and based on that knowledge will know how to support learners' during the learning process. Table 4.3 shows the weights of each feature's weights in predicting the final grades, whereas the plot in Figure 4.20 displays the importance of features in graphical form.

Table 4.3: Independent features weights in predicting the final grades

Feature	Feature Weight	Feature	Feature Weight
MP3	0.345631	NPS	0.010292
MP1	0.264391	NTRA	0.009293
MP2	0.246556	APV	0.009143
NTAQ	0.051794	SPV	0.007157
AST	0.018429	TRR	0.006407
ODGP	0.013156	NVRA	0.005492
NPP	0.012259		

- Learners and instructors can get a better understanding of m-learning model logic, accuracy, interpretation, and can improve it by focusing only on important features.

- The performance of the m-learning model can be increased by removing those features that are not significant in the construction of the model.
- Important features can also help us in making the m-learning model more interpretable. For example, the interpretable m-learning model can tell learners not only about their performance, but also which features contribute most during their learning process.

As discussed, the Random forest combines different decision trees to know about the essential features that drive the model, i.e., in any given model, these features are most important in explaining the relationship with the target feature. The decision tree assigns a node to each feature, and it split values based on a relational condition (greater than, less than, equal to, not equal to, greater than equal to, less than equal to) so that similar values end up in the same set after the split. The node's condition is based on impurity, which is how much information is gained (entropy) during model training. The important features are those that contribute to decreasing the overall impurity. Looking at the graph in figure 4.20, we can conclude that the top four important features are MP3, MP2, MP1, and NTAQ.

4.4.1 Permutation Feature Importance

Features permutation measures the importance of features by observing how features reshuffling or re-arrangement influence the model's performance and accuracy. We first train the baseline model with the default feature arrangement and record the accuracy score using this approach. Next, we reshuffle the values from one feature in our dataset and pass the dataset again to the model to obtain the model accuracy of the modified dataset. The feature importance of the reshuffle feature values is the difference between the initial benchmark score and the modified dataset. The values reshuffling was performed for all the features in the dataset. The features permutation approach is a reasonably efficient and reliable technique that can be applied to any model. On the other hand, it may be computationally expensive than the default feature importance approach. One other disadvantage of this approach is that it overestimates the importance

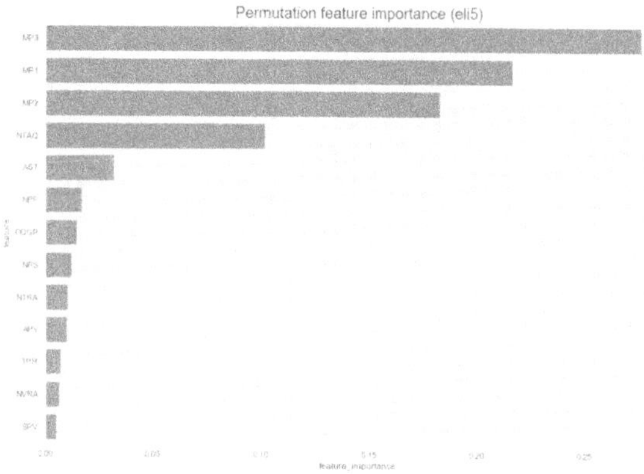

Figure 4.21: Permutation feature importance measurement using eli5 Python library

of correlated features. The plot in figure 4.21 confirms that the top four predictable features remain in the same position; however, their importance scores are different, particularly for the NTAQ feature.

4.4.2 Drop Column Feature Importance

Using this approach, we determined the importance of a feature by comparing the m-learning model with all features versus a model with all features except that feature. This approach provides a more accurate feature importance score but may incur a potentially high computation cost due to comparing two models with variant datasets. The features shown as negative importance in figure 4.22 are those features that will improve the model performance after removing from the dataset set. In our case, we have two features APV and NTRA, that exhibit this behavior. Moreover, MP1, MP2, and MP3 are no longer important predictable features and drop below the graph. Lastly, AST, NTAQ, and NPP now have the highest importance in predicting the learners' performance scores.

4.5 Experimental Design: Performing Early Intervention Experiment During Learning Process

In the last stage of our research, we conducted an early intervention experiment to figure out whether pre-emptive measures influence learners in improving their study performance. The experiment was carried out on those learners who achieved perfor-

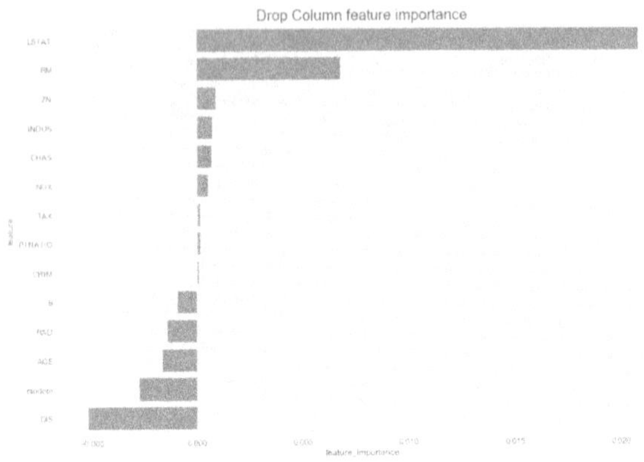

Figure 4.22: Drop column feature importance (dropping modules performance scores)

mance grades D and F (learners having 11 or less than 11 scores in their final performance quiz). At this point, our m-learning model is trained with learners' features and is matured enough to predict the learning performance of existing and new learners. The goal of carrying out an early intervention experiment was to motivate and encourage weak learners to improve their learning performance and make m-learning model predictions false. A total of 425 learners (having D and F grades) were selected and were divided into control and experimental groups (the control group having 212 learners whereas the experimental group having 213 learners). Control group learners were independent of pre-emptive measures, early intervention and received normal learning content, guidance, and quizzes, whereas experimental group learners received motivational triggers, adaptive triggers, and adaptive learning content. Moreover, adaptive navigational paths were recommended to experimental group learners during their learning process. The early intervention experiment lasted for 50 days. Below we discuss learning content types and how adaptive triggers are related to them.

- Learning content: As discussed earlier two types of learning content were used in the learnit app i.e. text and video.
- Quizzes. Each course consists of three quizzes and a final quiz was conducted after completion of three courses.
- Online discussion forum. An online discussion forum was provided to learners where after each learning activity, the learners can post problems related to that

activity.

During the early intervention experiment, different adaptive triggers were presented to experimental group learners during their learning process. The adaptive triggers were presented according to learners' preferred learning content types and how adaptive triggers are assigned to learning content. Below in table 4.4 we give some examples of adaptive triggers, their timings, and how they are related to learning content types.

Table 4.4: Adaptive Triggers along with their timings

S.No	Adaptive triggers	Timings
T1	It is better to write/note important points on hard paper during the learning process. It will help you in interpreting the same learning topic during revision.	Before
T2	Make sure that you have revised the topic and have gained a clear understanding of it.	After
T3	Writing a summary of the topic will help you in future references.	After
T4	Please make sure that you have not missed any point and have covered all the key points related to the current topic.	After
T5	Share your understanding of the current topic with your colleagues on the online discussion group.	Before/After
T6	If you have gained a complete understanding of the current topic then do visit the online discussion group and help your colleagues having problems with the same learning topic.	After
T7	Share your problem and request the assigned instructor to help you if you haven't understood the current learning topic.	After
T8	Challenge yourself and attempt the quiz if you have mastered the current learning topic.	After
T9	Note down important points and problems linked to the current learning topic and discuss them with your colleagues and instructor.	Before/After
T10	Improve your current quiz score to get a complete understanding of the learning topic.	After

Next, we list the assignment of adaptive triggers to learning content.

Next, we list the assignment of adaptive triggers to learning content.

- Text (T1-T10)
- Video (T4, T5, T6, T7, T8, T9)
- Quizzes (T5, T6, T7, T8, T9)
- Online discussion forum (T8, T9, T10)

The timings of the adaptive triggers are before and after. Before triggers refer to those triggers that are presented to learners before they have started the learning activity whereas after triggers means presenting triggers to learners after they have finished a learning activity. The purpose of adaptive triggers was to assist learners during their learning process and help them in understanding exhaustive and difficult topics.

Likewise, different motivational triggers were presented to learners on their smartphones according to their performance scores. Table 4.5 shows learners' performance state, the assigned motivational triggers, and their examples.

Table 4.5: Learners performance state, type of motivational triggers, and their example

Learners' Performance State	Type of motivational trigger	Motivational trigger example
Below Average Learners (Learners having low-performance score)	Fear	Continues poor performance will result in your relegation/ceased state
	Hope	Consistent 3 hours of study per day will make you one of the best learners of the class
	Suggestion	Refers to the class group for newly uploaded video by the class instructor
Average Learners (improving learners)	Praise	Weldon and congratulation. You now are in the top 5 learners of your class
	Appreciation	Your improving academic performance was praised by instructors in the meeting. Keep it up.
	Social Acceptance	Imran Ahmed is now among the top 5 learners (Trigger sent to all class learners)
Good Consistent Learners	Reward	The chairman will award you the best performance certificate for showing consistent performance.

Table 4.5 continued from previous page

	Praise	You showed excellent performance in the quiz. Weldon.
	Appreciation	All the instructors and chairman appreciate your consistent performance throughout the semester.

Table 4.6: End-User Computing Satisfaction (EUCS) questionnaire result.

	Dimensions	Questions	Mean Score
1	Usefulness	I contemplate the use of the learnit app enhanced my programming skills.	4.68
2	Usefulness	I think the learnit app helped me a lot in understanding computer programming.	4.59
3	Ease-of-use	The use of the learnit app was very easy.	4.52
4	Ease-of-use	I think learners can use the learnit app without the instructor's help.	4.58
5	Engaging	I was interested in using the learnit app for enhancing my programming skills.	4.78
6	Engaging	Learnit app was able to involve me in learning computer programming daily.	4.71
7	Timeliness	The learning content was provided on time by the learnit app.	4.34
8	Timeliness	Learnit app knew when and what to send to learners while using it.	4.53
9	Adaptive	The learning content provided by the learnit app was tailored and adaptive.	4.78
10	Adaptive	Learnit app enabled me to learn computer programming at my own pace.	4.23
11	Attitude towards using the application	I will use similar types of applications in the future.	4.55
12	Attitude towards using the application	I will continue to use the learnit app to enhance my programming skills.	4.34

BJ Fogg, in his Fogg Behavior Model (FBM), suggests that three elements, i.e., ability, motivation, and the trigger (prompt or message), must be present at the same time for a learning behavior to occur [170]. When improvement does not happen during the learning process, at least one of the three elements is missing. Figure 4.23 shows the pictorial representation of FBM along with its three elements. Adaptive triggers aimed to help learners while they study, whereas motivational triggers aimed to increase learners' motivation towards learning. One of the essential factors related to adaptive

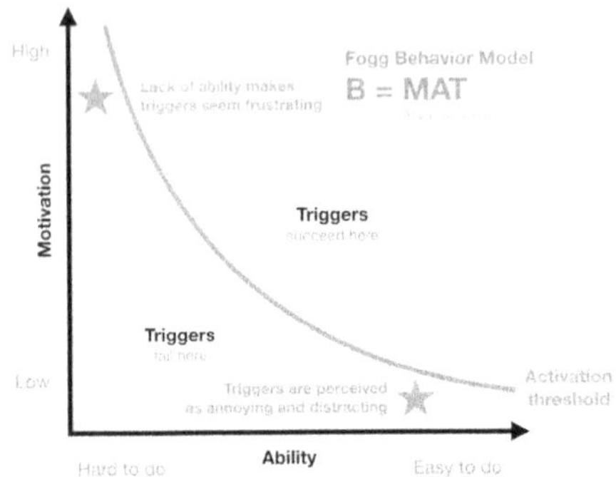

Figure 4.23: Fogg Behavior Model and its elements.

Figure 4.24: Performance comparison of the experimental and control groups after early intervention experiment.

and motivational triggers is opportune time. Opportune time is the right time at which adaptive and motivational triggers are sent to the learners on their smartphones. An example of sending motivational triggers at the opportune time is right after the learner quiz activity if the learner's performance in that quiz is not good. Similarly, praise and appreciation triggers were sent to the good consistent learners once a week as too many motivational triggers may overwhelm or annoy learners during the learning process.

4.6 Data Analysis: Early Intervention Experiment Results Discussion

After 50 days, we analyzed the final score results of the experimental and control group. Figure 4.24 shows a line graph representing both the experimental and control groups' final scores. The line graph confirms that the intervened learners showed overall better performance than un-intervened learners in the final quiz. Moreover, the statistical analysis result revealed that intervened learners showed an overall 12.30 percent higher performance than un-intervened learners. These results confirm that proactive or preemptive measures affect learners' behavior and improve their learning performance scores.

4.6.1 Early Intervention Experiment Discussion

Our m-learning model successfully identified different learning behavior for diverse learners by identifying key features and how they affect learners' learning performance. Furthermore, early intervention experiment results revealed that through adaptive and tailored guidance, learners' learning behavior could be enhanced and improved. Once trained with enough data, the m-learning model can identify new and old learners and how their learning features affect their study performance. Diverse learners have strengths and weaknesses represented by the learning features. Not only the m-learning model can identify which features affect study performance, but it can also determine which features are the most important ones (features having higher weights) in improving the learning performance of learners. A more adaptive and tailored application could be developed if it considers learning features along with their weights.

While predicting the performance of new learners, our m-learning model will not have their features information however, it has learned from the features of previous learners and can make predictions for new learners. When new learners' features and their instances are fed into the m-learning model, the model could determine feature weights and how they affect the final performance. Subsequently, preemptive measures could be taken for those learners who are unlikely to show good performance in the future.

4.6.2 Analyzing Learners' satisfaction using the EUCS model

To elicit learners' experience using the learnit application, the End-User Computing Satisfaction (EUCS) model was used. As discussed in the proposed system chapter, the learnit application is based on the m-learning model, which uses an artificial neural network (ANN) to determine how independent learning features affect learners' final learning performance. The EUCS model is composed of a questionnaire that can measure users' attitudes towards using a computer application. With little modification in the EUCS questionnaire, an online survey was steered on 213 experimental group learners. the EUCS questionnaire measured 6 dimensions of learnit application: usefulness, ease-of-use, engagement, timeliness, adaptiveness, and attitude towards learnit application. The online survey comprised of 12 questions covering the 6 dimensions of learnit application. The 6 dimensions were measured on 5 points Likert-scale where 5 represents "strongly agree", 4 represents "agree", 3 represents "neutral," 2 represents "disagree," and 1 represents the "strongly disagree" option. During the experiment, the mean learner satisfaction was set to greater than 4, which infer that the learners were satisfied while they used the learnit application. Table 4.6 shows the 6 dimensions, the related questions, and learners' satisfaction mean score.

The response to dimensions 1 and 2 (usefulness) indicates that the learnit app was successful in increasing the job ability i.e. programming skills of learners (m= 4.68, m = 4.59). The response to dimensions 3 and 4 shows that the learnit app interaction with learners was very simple and user-friendly (m= 4.52, m = 4.58). The response to dimensions 5 and 6 indicates that the learnit app was able to draw the attention of learners in giving time to learn computer programming using mobile devices (m= 4.78, m = 4.71). The response to dimensions 7 and 8 implies that the learnit app considered the experimental group preferred learning time and sent programming content accordingly (m= 4.34, m = 4.53). The learner replies to dimensions 9 and 10 revealed that tailored programming content was delivered to learners according to their learning abilities and performance by the learnit app (m= 4.78, m = 4.23). The answer to dimensions 11 and 12 reflects that learners agreed to use a learnit or a similar type of application in the future to increase their programming skills (m= 4.55, m = 4.34).

In chapter 4, we worked on important objectives mentioned in chapter 1. First, an M-learning model was trained using Artificial Neural Network (ANN). A correlation matrix was generated for features analysis to reveal the correlation among features and correlation between independent and dependent features. From the correlation matrix, we observed that MP1, MP2, MP3, ODGP, NPP, and NTRA were the significant features that affect M-learners' overall performance. A technique called back-propagation was used to minimize the M-learner model loss, which reduced the difference between predicted and actual output result. To determine each independent feature's importance in predicting the final performance grades, the Random Forest (RF) ensemble method was used. Using the RF ensemble method, it was revealed that among 13 independent features, MP1, MP2, MP3, and NTAQ were the top 4 important features that contributed significantly to increasing M-learners' final performance. Furthermore, a technique called permutation feature importance was used to determine how features re-shuffling influences the performance and accuracy of the model. After applying the permutation feature importance technique, the four independent features remain in the same position; however, their importance scores were different, particularly for the NTAQ feature. An early intervention experiment was performed after the M-learner used ILS to determine whether pre-emptive measures influence the learners' study performance. Finally, the EUCS model was used to analyze learners' satisfaction towards using ILS. The EUCS model results revealed that overall, learners were satisfied with using ILS and they agree to use a similar type of system in the future to improve their study performance.

CHAPTER 5: CONCLUSION AND FUTURE WORK

This research work presents the m-learning model generated by an Artificial Neural Network (ANN) from learners' weighted features. Established on the m-learning model, a prototype Android app called learnit was developed and tested on m-learners for 50 days. The m-learning model can be used by those applications or systems that need to make learning self-paced, engaging, and adaptive. The artificial neural network (ANN) generates a tailored m-learning model for each learner according to his/her weighted learning features.

The proposed m-learning model can classify learners in different performance groups with high accuracy and can provide tailored guidance and adaptive learning content to the learners. Initially, m-learners' features are elicited, and the m-learners profile is defined on the online Firebase cloud while interacting with the learnit application. The m-learners features are preprocessed, encoded, scaled, and divided into training and testing sets before delivering them to the ANN algorithm. After the features preprocessing step, the m-learners' features are supplied to the ANN for m-learners classification and features weights tuning. For efficient processing, forward propagation and the matrix multiplication technique were used to transmit multiple learners' records at a time through ANN. To minimize the output error and increase m-learning model accuracy, the ANN was further trained using the back-propagation technique. For increasing m-learning model accuracy, suitable synapses weights must be defined across ANN architecture. Stochastic gradient descent was used for assigning appropriate weights across ANN synapses. Stochastic gradient descent also makes sure that less computation and memory resources are consumed while training ANN. The ANN training process establishes a relationship between independent features and dependent features (learners' final performance). This relationship also defines independent feature weights and how they contribute to increasing the learners' final performance. The relationship between independent features and the dependent feature is established for every learner; thus, every learner has a unique m-learning model representing the individual learner's learning behavior.

After completing the m-learning model training process, we tested its accuracy by removing some essential features. We assumed that module performance features (MP1, MP2, and MP3) contribute significantly to increasing learners' final performance; therefore, the testing process was conducted on four sub-models, namely m-learning model 1 m-learning model 2, m-learning model 3, and m-learning model 4. M-learning model 1 was comprised of all the features except final performance, m-learning model 2 was like m-learning model 1 but without module performance 3 (MP3) feature. Similarly, m-learning model 3 was like m-learning model 2 without the module performance 2 (MP2) feature. Lastly, m-learning model 4 was similar to m-learning model 3 but without module performance 1 (MP1) feature. M-learners were classified using these four models, and subsequently, the models achieved an accuracy of 90.77, 87.69, 83.85, and 80.00 percent, respectively. We observed that as modules' performance scores are removed from the independent features dataset, the models' accuracy is decreased. This experiment revealed that module performance scores are vital in predicting the final performance score. Moreover, using the Random Forest ensemble method, we determined the weights of all independent features. The feature weights indicate how important each feature is in contributing to the learner's final performance. Using the eli5 Python library, an experiment was conducted to determine the most important features when module performance score features (MP1, MP2, MP3) are ignored. The results revealed that Average Study Time (AST), Number of Times Attempted Quiz (NTAQ), and Number of Problems Posted (NPP) features become important features in predicting the final performance score when module performance features (MP1, MP2, MP3) are ignored. The feature weights and importance results helped in making the m-learning model adaptive and learners oriented. Based on the m-learning model, a learnit app was developed and tested to know how much adaptiveness in the learning process helps learners increase their study performance.

In the last stage of our research, an early intervention experiment was carried out on those learners who achieved D and F grades in their final performance. The early intervention experiment aimed to reveal whether early interference or intervention during the learning process improves study performance. M-learners were divided into two groups, namely the control group and the experimental group. The control group comprised 212 learners and was independent of the intervention, and received normal learning content. Experimental groups included 213 learners and received adaptive content, guidance, help, and support during their learning process. Moreover, two types of triggers (messages), namely adaptive and motivational triggers, were presented to experimental groups participants on their mobile devices. The purpose of adaptive triggers was to increase study performance and provide adaptive learning content to m-learners, whereas the purpose of motivational triggers was to encourage/motivate m-learners towards improving their learning performance. The early intervention experiment revealed that experimental group performance increased by 12.30 percent compared to the control group. The early intervention experiment showed that pro-active or early measures could increase learners' learning output.

Lastly, the End-User Computing Satisfaction (EUCS) model was used to elicit m-learners satisfaction towards using the learnit application. An online survey was conducted on 213 experimental group participants, and six dimensions of learnit application, namely usefulness, ease-of-use, timeliness, engaging, adaptiveness, and attitude of learners towards using learnit application, were measured. The survey questionnaire consisted of 12 questions covering the six dimensions of the learnit application. The mean satisfaction score was greater than 4 for all dimensions, which concluded that the m-learners showed their satisfaction towards using the learnit application.

In the future, we would like to test the m-learning model on a larger scale and integrate it with the Learning Management System (LMS). By doing so, the m-learning model will help us in making LMS an adaptive and more user-friendly system. One of the disadvantages of traditional LMS systems is that they are based on a one-size-fits-all approach where all learners are treated equally, not considering individual weaknesses, strengths, preferences, and personalized learning behavior. By integrating the m-learning model with LMS, we might make the learning process self-paced and adaptive. Moreover, we would also like to improve approaches presented in this research study by combining other deep neural network algorithms such as Recurrent Neural Network (RNN), Deep Autoencoders, Self-Organizing Maps (SOMs), and Long Short-Term Memory (LSTM) with the proposed m-learning model.

References

[1] I. H. Sarker and A. Kayes, "Abc-ruleminer: User behavioral rule-based machine learning method for context-aware intelligent services," *Journal of Network and Computer Applications*, vol. 168, p. 102762, 2020.

[2] N. Petkov, N. Strisciuglio, and C. M. Travieso-González, *Applications of Intelligent Systems: Proceedings of the 1st International APPIS Conference 2018.* IOS Press, 2018, vol. 310.

[3] G. Yalamov, "Possible negative impacts of intelligent learning systems on the development of the user's identity," in *International Conference on the Development of Education in Eurasia (ICDEE 2019).* Atlantis Press, 2019, pp. 117–119.

[4] M. A. Al-Nakhal and S. S. Abu-Naser, "Adaptive intelligent tutoring system for learning computer theory," 2017.

[5] S. Ennouamani and Z. Mahani, "An overview of adaptive e-learning systems," in *2017 eighth international conference on intelligent computing and information systems (ICICIS).* IEEE, 2017, pp. 342–347.

[6] M. M. Hilles and S. S. A. Naser, "Knowledge-based intelligent tutoring system for teaching mongo database," 2017.

[7] M. Ahmed, N. A. Sangi, and A. Mahmood, "A model of adaptive e-learning in an odl environment," *Mehran University Research Journal Of Engineering & Technology*, 2018.

[8] A. Kumar and N. J. Ahuja, "An adaptive framework of learner model using learner characteristics for intelligent tutoring systems," in *Intelligent communication, control and devices.* Springer, 2020, pp. 425–433.

[9] R. A. Sottilare, A. C. Graesser, X. Hu, A. Olney, B. Nye, and A. M. Sinatra, *Design Recommendations for Intelligent Tutoring Systems: Volume 4-Domain Modeling.* US Army Research Laboratory, 2016, vol. 4.

[10] P. A. Akiki, A. K. Bandara, and Y. Yu, "Adaptive model-driven user interface development systems," *ACM Computing Surveys (CSUR)*, vol. 47, no. 1, pp. 1–33, 2014.

[11] N. Zanjani, "The important elements of lms design that affect user engagement with e-learning tools within lmss in the higher education sector," *Australasian Journal of Educational Technology*, vol. 33, no. 1, 2017.

[12] Y. Jung and J. Lee, "Learning engagement and persistence in massive open online courses (moocs)," *Computers & Education*, vol. 122, pp. 9–22, 2018.

[13] T. Rohloff, D. Sauer, and C. Meinel, "On the acceptance and usefulness of personalized learning objectives in moocs," in *Proceedings of the Sixth (2019) ACM Conference on Learning@ Scale*, 2019, pp. 1–10.

[14] M. Liu, E. McKelroy, S. B. Corliss, and J. Carrigan, "Investigating the effect of an adaptive learning intervention on students' learning," *Educational technology research and development*, vol. 65, no. 6, pp. 1605–1625, 2017.

[15] M. Vandewaetere, P. Desmet, and G. Clarebout, "The contribution of learner characteristics in the development of computer-based adaptive learning environments," *Computers in Human Behavior*, vol. 27, no. 1, pp. 118–130, 2011.

[16] H. Hamidi and A. Chavoshi, "Analysis of the essential factors for the adoption of mobile learning in higher education: A case study of students of the university of technology," *Telematics and Informatics*, vol. 35, no. 4, pp. 1053–1070, 2018.

[17] M. M. Elaish, L. Shuib, N. A. Ghani, E. Yadegaridehkordi, and M. Alaa, "Mobile learning for english language acquisition: taxonomy, challenges, and recommendations," *Ieee Access*, vol. 5, pp. 19 033–19 047, 2017.

[18] M. Alrasheedi and L. F. Capretz, "Determination of critical success factors affecting mobile learning: A meta-analysis approach," *arXiv preprint arXiv:1801.04288*, 2018.

[19] X.-W. Pan, L. Ding, X.-Y. Zhu, and Z.-X. Yang, "A social approach to high-level context generation for supporting context-aware m-learning," *Eurasia Journal of*

Mathematics, Science and Technology Education, vol. 13, no. 7, pp. 3675–3686, 2017.

[20] H. Li, H. Li, S. Zhang, Z. Zhong, and J. Cheng, "Intelligent learning system based on personalized recommendation technology," *Neural Computing and Applications*, vol. 31, no. 9, pp. 4455–4462, 2019.

[21] N. B. A. Normadhi, L. Shuib, H. N. M. Nasir, A. Bimba, N. Idris, and V. Balakrishnan, "Identification of personal traits in adaptive learning environment: Systematic literature review," *Computers & Education*, vol. 130, pp. 168–190, 2019.

[22] B. Vesin, K. Mangaroska, and M. Giannakos, "Learning in smart environments: user-centered design and analytics of an adaptive learning system," *Smart Learning Environments*, vol. 5, no. 1, pp. 1–21, 2018.

[23] B. J. Evans, R. B. Baker, and T. S. Dee, "Persistence patterns in massive open online courses (moocs)," *The Journal of Higher Education*, vol. 87, no. 2, pp. 206–242, 2016.

[24] S. Amershi and C. Conati, "Unsupervised and supervised machine learning in user modeling for intelligent learning environments," in *Proceedings of the 12th international conference on Intelligent user interfaces*, 2007, pp. 72–81.

[25] A. Lin, Q. Wu, A. A. Heidari, Y. Xu, H. Chen, W. Geng, C. Li *et al.*, "Predicting intentions of students for master programs using a chaos-induced sine cosine-based fuzzy k-nearest neighbor classifier," *Ieee Access*, vol. 7, pp. 67 235–67 248, 2019.

[26] J. R. Anderson, S. Betts, J. L. Ferris, and J. M. Fincham, "Neural imaging to track mental states while using an intelligent tutoring system," *Proceedings of the National Academy of Sciences*, vol. 107, no. 15, pp. 7018–7023, 2010.

[27] B. Drăgulescu, M. Bucos, and R. Vasiu, "Predicting assignment submissions in a multi-class classification problem," *TEM Journal*, vol. 4, no. 3, p. 244, 2015.

[28] W. Hämäläinen and M. Vinni, "Comparison of machine learning methods for intelligent tutoring systems," in *International conference on intelligent tutoring systems*. Springer, 2006, pp. 525–534.

[29] J. Makhlouf and T. Mine, "Analysis of click-stream data to predict stem careers from student usage of an intelligent tutoring system." *Journal of Educational Data Mining*, vol. 12, no. 2, pp. 1–18, 2020.

[30] D. Durães, R. Toala, F. Gonçalves, and P. Novais, "Intelligent tutoring system to improve learning outcomes," *AI Communications*, vol. 32, no. 3, pp. 161–174, 2019.

[31] J. Melesko and E. Kurilovas, "Semantic technologies in e-learning: Learning analytics and artificial neural networks in personalised learning systems," in *Proceedings of the 8th International Conference on Web Intelligence, Mining and Semantics*, 2018, pp. 1–7.

[32] G. Carleo, I. Cirac, K. Cranmer, L. Daudet, M. Schuld, N. Tishby, L. Vogt-Maranto, and L. Zdeborová, "Machine learning and the physical sciences," *Reviews of Modern Physics*, vol. 91, no. 4, p. 045002, 2019.

[33] J. Cai, J. Luo, S. Wang, and S. Yang, "Feature selection in machine learning: A new perspective," *Neurocomputing*, vol. 300, pp. 70–79, 2018.

[34] P. Lavanya, K. Kouer, and S. Mallappa, "Effective feature representation using symbolic approach for classification and clustering of big data," *Expert Systems with Applications*, p. 114658, 2021.

[35] S. Pouyanfar, S. Sadiq, Y. Yan, H. Tian, Y. Tao, M. P. Reyes, M.-L. Shyu, S.-C. Chen, and S. Iyengar, "A survey on deep learning: Algorithms, techniques, and applications," *ACM Computing Surveys (CSUR)*, vol. 51, no. 5, pp. 1–36, 2018.

[36] H. Liang, X. Sun, Y. Sun, and Y. Gao, "Text feature extraction based on deep learning: a review," *EURASIP journal on wireless communications and networking*, vol. 2017, no. 1, pp. 1–12, 2017.

[37] M. Akçakaya, S. Moeller, S. Weingärtner, and K. Uğurbil, "Scan-specific robust artificial-neural-networks for k-space interpolation (raki) reconstruction: Database-free deep learning for fast imaging," *Magnetic resonance in medicine*, vol. 81, no. 1, pp. 439–453, 2019.

[38] S. Albawi, T. A. Mohammed, and S. Al-Zawi, "Understanding of a convolutional neural network," in *2017 International Conference on Engineering and Technology (ICET)*. Ieee, 2017, pp. 1–6.

[39] T. Donkers, B. Loepp, and J. Ziegler, "Sequential user-based recurrent neural network recommendations," in *Proceedings of the Eleventh ACM Conference on Recommender Systems*, 2017, pp. 152–160.

[40] D. Miljković, "Brief review of self-organizing maps," in *2017 40th International Convention on Information and Communication Technology, Electronics and Microelectronics (MIPRO)*. IEEE, 2017, pp. 1061–1066.

[41] A. Hernández-Blanco, B. Herrera-Flores, D. Tomás, and B. Navarro-Colorado, "A systematic review of deep learning approaches to educational data mining," *Complexity*, vol. 2019, 2019.

[42] N. Bosch and L. Paquette, "Unsupervised deep autoencoders for feature extraction with educational data," in *Deep Learning with Educational Data Workshop at the 10th International Conference on Educational Data Mining*, 2017.

[43] C. Perrotta and N. Selwyn, "Deep learning goes to school: Toward a relational understanding of ai in education," *Learning, Media and Technology*, vol. 45, no. 3, pp. 251–269, 2020.

[44] H. Crompton and D. Burke, "The use of mobile learning in higher education: A systematic review," *Computers & Education*, vol. 123, pp. 53–64, 2018.

[45] F. E. Louhab, A. Bahnasse, and M. Talea, "Towards an adaptive formative assessment in context-aware mobile learning," *Procedia Computer Science*, vol. 135, pp. 441–448, 2018.

[46] N. Kara and K. Cagiltay, "In-service preschool teachers' thoughts about technology and technology use in early educational settings," *Contemporary Educational Technology*, vol. 8, no. 2, pp. 119–141, 2017.

[47] G. D. Wagner and D. D. Flannery, "A quantitative study of factors affecting learner acceptance of a computer-based training support tool," *Journal of European Industrial Training*, 2004.

[48] C. W. Coley, W. H. Green, and K. F. Jensen, "Machine learning in computer-aided synthesis planning," *Accounts of chemical research*, vol. 51, no. 5, pp. 1281–1289, 2018.

[49] A. B. Wilson, K. M. Brown, J. Misch, C. H. Miller, B. A. Klein, M. A. Taylor, M. Goodwin, E. K. Boyle, C. Hoppe, and M. D. Lazarus, "Breaking with tradition: A scoping meta-analysis analyzing the effects of student-centered learning and computer-aided instruction on student performance in anatomy," *Anatomical sciences education*, vol. 12, no. 1, pp. 61–73, 2019.

[50] N. J. Kim, B. R. Belland, and A. E. Walker, "Effectiveness of computer-based scaffolding in the context of problem-based learning for stem education: Bayesian meta-analysis," *Educational Psychology Review*, vol. 30, no. 2, pp. 397–429, 2018.

[51] M. A. Ozer, F. Govsa, and A. H. Bati, "Web-based teaching video packages on anatomical education," *Surgical and Radiologic Anatomy*, vol. 39, no. 11, pp. 1253–1261, 2017.

[52] R. Deng, P. Benckendorff, and D. Gannaway, "Progress and new directions for teaching and learning in moocs," *Computers & Education*, vol. 129, pp. 48–60, 2019.

[53] J. Traxler, "Distance learning—predictions and possibilities," *Education Sciences*, vol. 8, no. 1, p. 35, 2018.

[54] J. Gardner and C. Brooks, "Student success prediction in moocs," *User Modeling and User-Adapted Interaction*, vol. 28, no. 2, pp. 127–203, 2018.

[55] S. Kumar Basak, M. Wotto, and P. Belanger, "E-learning, m-learning and d-learning: Conceptual definition and comparative analysis," *E-Learning and Digital Media*, vol. 15, no. 4, pp. 191–216, 2018.

[56] V. Arkorful and N. Abaidoo, "The role of e-learning, advantages and disadvantages of its adoption in higher education," *International Journal of Instructional Technology and Distance Learning*, vol. 12, no. 1, pp. 29–42, 2015.

[57] B. Gros and F. J. García-Peñalvo, "Future trends in the design strategies and technological affordances of e-learning." Springer, 2016.

[58] S. Yu, M. Ally, and A. Tsinakos, *Mobile and ubiquitous learning: An international handbook.* Springer, 2017.

[59] J. Gikas and M. M. Grant, "Mobile computing devices in higher education: Student perspectives on learning with cellphones, smartphones & social media," *The Internet and Higher Education*, vol. 19, pp. 18–26, 2013.

[60] M. R. Rahimi, J. Ren, C. H. Liu, A. V. Vasilakos, and N. Venkatasubramanian, "Mobile cloud computing: A survey, state of art and future directions," *Mobile Networks and Applications*, vol. 19, no. 2, pp. 133–143, 2014.

[61] J. Fombona, Á. Pascual-Sevillana, and M. Gonzalez-Videgaray, "M-learning and augmented reality: A review of the scientific literature on the wos repository," *Comunicar. Media Education Research Journal*, vol. 25, no. 2, 2017.

[62] T. Cochrane, "A summary and critique of mlearning research and practice," *Handbook of mobile learning*, pp. 24–34, 2013.

[63] O. T. Murray and N. R. Olcese, "Teaching and learning with ipads, ready or not?" *TechTrends*, vol. 55, no. 6, pp. 42–48, 2011.

[64] R. Robinson, *Digital Thinking and Mobile Teaching.* bookboon. com, 2017.

[65] D. W. Fahlman, "Reflections on distributive leadership for work-based mobile learning of canadian registered nurses," *International Journal of Mobile and Blended Learning (IJMBL)*, vol. 9, no. 3, pp. 1–11, 2017.

[66] S. Gabrielli, S. Kimani, and T. Catarci, "The design of microlearning experiences: A research agenda (on microlearning)," 2017.

[67] A. F. M. Ayub, S. H. Zaini, W. S. Luan, and W. M. W. Jaafar, "The influence of mobile self-efficacy, personal innovativeness and readiness towards students' attitudes towards the use of mobile apps in learning and teaching," *International Journal of Academic Research in Business and Social Sciences*, vol. 7, no. 14, pp. 364–374, 2017.

[68] S. Schuck, M. Kearney, and K. Burden, "Exploring mobile learning in the third space," *Technology, Pedagogy and Education*, vol. 26, no. 2, pp. 121–137, 2017.

[69] F. Trede, L. Markauskaite, C. McEwen, and S. Macfarlane, "mlearning in workplace learning," in *Education for Practice in a Hybrid Space*. Springer, 2019, pp. 33–46.

[70] M. Al-Emran and S. A. Salloum, "Students' attitudes towards the use of mobile technologies in e-evaluation." *Int. J. Interact. Mob. Technol.*, vol. 11, no. 5, pp. 195–202, 2017.

[71] M. Makoe and T. Shandu, "Developing a mobile app for learning english vocabulary in an open distance learning context," *International Review of Research in Open and Distributed Learning*, vol. 19, no. 4, 2018.

[72] G. Forecast, "Cisco visual networking index: global mobile data traffic forecast update, 2017–2022," *Update*, vol. 2017, p. 2022, 2019.

[73] E. M. Fox, "Mobile technology: a tool to increase global competency among higher education students," *International Review of Research in Open and Distributed Learning*, vol. 20, no. 2, 2019.

[74] H. F. Hanafi, N. A. Zainuddin, M. H. Abd Wahab, and A. H. Ariffin, "Technology acceptance of a novel mobile learning application among university undergraduates," *International Business Education Journal*, vol. 11, no. 1, pp. 16–24, 2018.

[75] A. Peramunugamage, H. Ratnayake, and S. Karunanayaka, "Work-in-progress: Development of a framework to foster collaborative learning among engineering students using moodle mobile app," in *International Conference on Interactive Collaborative Learning*. Springer, 2018, pp. 3–13.

[76] K. Akherfi, M. Gerndt, and H. Harroud, "Mobile cloud computing for computation offloading: Issues and challenges," *Applied computing and informatics*, vol. 14, no. 1, pp. 1–16, 2018.

[77] S. S. Manvi and G. K. Shyam, "Resource management for infrastructure as a service (iaas) in cloud computing: A survey," *Journal of network and computer applications*, vol. 41, pp. 424–440, 2014.

[78] V. S. Sharma, S. Sengupta, and A. K. Mohamedrasheed, "Method and system for managing user state for applications deployed on platform as a service (paas) clouds," Apr. 25 2017, uS Patent 9,635,088.

[79] S. H. Kim, S. Y. Jang, and K. H. Yang, "Analysis of the determinants of software-as-a-service adoption in small businesses: Risks, benefits, and organizational and environmental factors," *Journal of Small Business Management*, vol. 55, no. 2, pp. 303–325, 2017.

[80] K. Haruna, M. Akmar Ismail, S. Suhendroyono, D. Damiasih, A. C. Pierewan, H. Chiroma, and T. Herawan, "Context-aware recommender system: A review of recent developmental process and future research direction," *Applied Sciences*, vol. 7, no. 12, p. 1211, 2017.

[81] S. van Engelenburg, M. Janssen, and B. Klievink, "Designing context-aware systems: A method for understanding and analysing context in practice," *Journal of logical and algebraic methods in programming*, vol. 103, pp. 79–104, 2019.

[82] A. Hasanov and T. H. Laine, "A survey of context-awareness in learning environments in 2010-2016." in *CSEDU (1)*, 2017, pp. 234–241.

[83] O. B. Sezer, E. Dogdu, and A. M. Ozbayoglu, "Context-aware computing, learning, and big data in internet of things: a survey," *IEEE Internet of Things Journal*, vol. 5, no. 1, pp. 1–27, 2017.

[84] M. A. Virtanen, E. Haavisto, E. Liikanen, and M. Kääriäinen, "Ubiquitous learning environments in higher education: A scoping literature review," *Education and Information Technologies*, vol. 23, no. 2, pp. 985–998, 2018.

[85] R. Madhubala and A. Akila, "Context aware and adaptive mobile learning: A survey," *Advances in Computational Sciences and Technology*, vol. 10, no. 5, pp. 1355–1370, 2017.

[86] S. Ennouamani, Z. Mahani, and L. Akharraz, "A context-aware mobile learning system for adapting learning content and format of presentation: design, validation and evaluation," *Education and Information Technologies*, pp. 1–37, 2020.

[87] R. Pelánek, "Bayesian knowledge tracing, logistic models, and beyond: an overview of learner modeling techniques," *User Modeling and User-Adapted Interaction*, vol. 27, no. 3, pp. 313–350, 2017.

[88] F. B. Baker and S.-H. Kim, *The basics of item response theory using R*. Springer, 2017.

[89] S. B. Kotsiantis, "Use of machine learning techniques for educational proposes: a decision support system for forecasting students' grades," *Artificial Intelligence Review*, vol. 37, no. 4, pp. 331–344, 2012.

[90] A. Talebi, J. Mahjoobi, M. T. Dastorani, and V. Moosavi, "Estimation of suspended sediment load using regression trees and model trees approaches (case study: Hyderabad drainage basin in iran)," *ISH Journal of Hydraulic Engineering*, vol. 23, no. 2, pp. 212–219, 2017.

[91] H. A. Isiyaka, A. Mustapha, H. Juahir, and P. Phil-Eze, "Water quality modelling using artificial neural network and multivariate statistical techniques," *Modeling Earth Systems and Environment*, vol. 5, no. 2, pp. 583–593, 2019.

[92] A. F. Schmidt and C. Finan, "Linear regression and the normality assumption," *Journal of clinical epidemiology*, vol. 98, pp. 146–151, 2018.

[93] I. Ahmad, M. Basheri, M. J. Iqbal, and A. Rahim, "Performance comparison of support vector machine, random forest, and extreme learning machine for intrusion detection," *IEEE access*, vol. 6, pp. 33 789–33 795, 2018.

[94] H. S. Alenezi and M. H. Faisal, "Utilizing crowdsourcing and machine learning in education: Literature review," *Education and Information Technologies*, pp. 1–16, 2020.

[95] A. Soni, V. Kumar, R. Kaur, and D. Hemavath, "Predicting student performance using data mining techniques," *International Journal of Pure and applied Mathematics*, vol. 119, no. 12, pp. 221–227, 2018.

[96] O. C. Santos and J. G. Boticario, "User-centred design and educational data mining support during the recommendations elicitation process in social online learning environments," *Expert Systems*, vol. 32, no. 2, pp. 293–311, 2015.

[97] T.-C. Hsia, A.-J. Shie, and L.-C. Chen, "Course planning of extension education to meet market demand by using data mining techniques–an example of chinkuo technology university in taiwan," *Expert Systems with Applications*, vol. 34, no. 1, pp. 596–602, 2008.

[98] J. P. Campbell, *Utilizing student data within the course management system to determine undergraduate student academic success: An exploratory study*. Purdue University, 2007.

[99] L. P. Macfadyen and S. Dawson, "Mining lms data to develop an "early warning system" for educators: A proof of concept," *Computers & education*, vol. 54, no. 2, pp. 588–599, 2010.

[100] F. Chiclana, R. Kumar, M. Mittal, M. Khari, J. M. Chatterjee, S. W. Baik *et al.*, "Arm–amo: an efficient association rule mining algorithm based on animal migration optimization," *Knowledge-Based Systems*, vol. 154, pp. 68–80, 2018.

[101] A. R. Anaya and J. G. Boticario, "Application of machine learning techniques to analyse student interactions and improve the collaboration process," *Expert Systems with Applications*, vol. 38, no. 2, pp. 1171–1181, 2011.

[102] W. Lin, S. Wang, Y. Wu, and T. Ye, "An empirical analysis on auto corporation training program planning by data mining techniques," *Expert Systems with Applications*, vol. 38, no. 5, pp. 5841–5850, 2011.

[103] L. K. Poon, S.-C. Kong, T. S. Yau, M. Wong, and M. H. Ling, "Learning analytics for monitoring students participation online: Visualizing navigational patterns on learning management system," in *International conference on blended learning*. Springer, 2017, pp. 166–176.

[104] S. Asadianfam, H. Kolivand, and S. Asadianfam, "A new approach for web usage mining using case based reasoning," *SN Applied Sciences*, vol. 2, no. 7, pp. 1–11, 2020.

[105] H. N. N. M. Thu, "Discovering generalized association rule in web usage mining by frequent pattern tree (fp-tree)," Ph.D. dissertation, University of Computer Studies, Yangon, 2019.

[106] I. Goodfellow, Y. Bengio, A. Courville, and Y. Bengio, *Deep learning*. MIT press Cambridge, 2016, vol. 1, no. 2.

[107] B. Guo, R. Zhang, G. Xu, C. Shi, and L. Yang, "Predicting students performance in educational data mining," in *2015 International Symposium on Educational Technology (ISET)*. IEEE, 2015, pp. 125–128.

[108] B. Akram, W. Min, E. Wiebe, B. Mott, K. E. Boyer, and J. Lester, "Improving stealth assessment in game-based learning with lstm-based analytics," in *International Conference on Educational Data Mining*, 2018.

[109] W. W. Fok, Y. He, H. A. Yeung, K. Law, K. Cheung, Y. Ai, and P. Ho, "Prediction model for students' future development by deep learning and tensorflow artificial intelligence engine," in *2018 4th international conference on information management (ICIM)*. IEEE, 2018, pp. 103–106.

[110] K. Abhinav, V. Subramanian, A. Dubey, P. Bhat, and A. D. Venkat, "Lecore: A framework for modeling learner's preference." in *EDM*, 2018.

[111] M. V. Amazona and A. A. Hernandez, "Modelling student performance using data mining techniques: Inputs for academic program development," in *Proceedings of the 2019 5th International Conference on Computing and Data Engineering*, 2019, pp. 36–40.

[112] J. G. A. Mantecon, H. A. Ghavidel, A. Zouaq, J. Jovanovic, and J. McDonald, "A comparison of features for the automatic labeling of student answers to open-ended questions." *International Educational Data Mining Society*, 2018.

[113] B.-H. Kim, E. Vizitei, and V. Ganapathi, "Gritnet: Student performance prediction with deep learning," *arXiv preprint arXiv:1804.07405*, 2018.

[114] M. Guo, E. Chou, D.-A. Huang, S. Song, S. Yeung, and L. Fei-Fei, "Neural graph matching networks for fewshot 3d action recognition," in *Proceedings of the European conference on computer vision (ECCV)*, 2018, pp. 653–669.

[115] S.-S. M. Ajibade, N. B. Ahmad, and S. M. Shamsuddin, "A data mining approach to predict academic performance of students using ensemble techniques,"

in *International Conference on Intelligent Systems Design and Applications*. Springer, 2018, pp. 749–760.

[116] M. M. Alam, K. Mohiuddin, A. K. Das, M. K. Islam, M. S. Kaonain, and M. H. Ali, "A reduced feature based neural network approach to classify the category of students," in *Proceedings of the 2nd International Conference on Innovation in Artificial Intelligence*, 2018, pp. 28–32.

[117] M. Khajah, R. V. Lindsey, and M. C. Mozer, "How deep is knowledge tracing?" *arXiv preprint arXiv:1604.02416*, 2016.

[118] X. Ma, Y. Yang, and Z. Zhou, "Using machine learning algorithm to predict student pass rates in online education," in *Proceedings of the 3rd International Conference on Multimedia Systems and Signal Processing*, 2018, pp. 156–161.

[119] A. Lalwani and S. Agrawal, "Few hundred parameters outperform few hundred thousand," in *Proceedings of the 10th International Conference on Educational Data Mining, EDM*, vol. 17. ERIC, 2017, pp. 448–453.

[120] M. El Fouki, N. Aknin *et al.*, "Multidimensional approach based on deep learning to improve the prediction performance of dnn models." *International Journal of Emerging Technologies in Learning*, vol. 14, no. 2, 2019.

[121] S. M. R. Abidi, M. Hussain, Y. Xu, and W. Zhang, "Prediction of confusion attempting algebra homework in an intelligent tutoring system through machine learning techniques for educational sustainable development," *Sustainability*, vol. 11, no. 1, p. 105, 2019.

[122] K. Hadullo, R. Oboko, and E. Omwenga, "Factors affecting asynchronous e-learning quality in developing countries university settings," *International Journal of Education and Development using ICT*, vol. 14, no. 1, 2018.

[123] W. Wang, H. Yu, and C. Miao, "Deep model for dropout prediction in moocs," in *Proceedings of the 2nd International Conference on Crowd Science and Engineering*, 2017, pp. 26–32.

[124] I. G. Ndukwe, B. K. Daniel, and R. J. Butson, "Data science approach for simulating educational data: towards the development of teaching outcome model (tom)," *Big Data and Cognitive Computing*, vol. 2, no. 3, p. 24, 2018.

[125] J. Whitehill, K. Mohan, D. Seaton, Y. Rosen, and D. Tingley, "Delving deeper into mooc student dropout prediction," *arXiv preprint arXiv:1702.06404*, 2017.

[126] Y. Chai, C.-U. Lei, X. Hu, and Y.-K. Kwok, "Wpss: dropout prediction for moocs using course progress normalization and subset selection," in *Proceedings of the Fifth Annual ACM Conference on Learning at Scale*, 2018, pp. 1–2.

[127] C.-K. Yeung and D.-Y. Yeung, "Addressing two problems in deep knowledge tracing via prediction-consistent regularization," in *Proceedings of the Fifth Annual ACM Conference on Learning at Scale*, 2018, pp. 1–10.

[128] S. Yanchao, Q. Xu, J. Minzheng, and B. Jing, "Research on undergraduate academic prediction model based on deep learning," in *2018 IEEE/ACIS 17th International Conference on Computer and Information Science (ICIS)*. IEEE, 2018, pp. 653–658.

[129] E. Tanuar, Y. Heryadi, B. S. Abbas, F. L. Gaol *et al.*, "Using machine learning techniques to earlier predict student's performance," in *2018 Indonesian Association for Pattern Recognition International Conference (INAPR)*. IEEE, 2018, pp. 85–89.

[130] J. Yang, K. Wang, X. Peng, and Y. Qiao, "Deep recurrent multi-instance learning with spatio-temporal features for engagement intensity prediction," in *Proceedings of the 20th ACM international conference on multimodal interaction*, 2018, pp. 594–598.

[131] D. Islam, "Analysis and assessment of skill using data mining," 2018.

[132] X. Wang, P. Wu, G. Liu, Q. Huang, X. Hu, and H. Xu, "Learning performance prediction via convolutional gru and explainable neural networks in e-learning environments," *Computing*, vol. 101, no. 6, pp. 587–604, 2019.

[133] A. A. Saa, M. Al-Emran, and K. Shaalan, "Mining student information system records to predict students' academic performance," in *International conference*

on advanced machine learning technologies and applications. Springer, 2019, pp. 229–239.

[134] A. Sales, A. Botelho, T. Patikorn, and N. T. Heffernan, "Using big data to sharpen design-based inference in a/b tests," in *Proceedings of the Eleventh International Conference on Educational Data Mining*, 2018.

[135] A. Sharma, A. Biswas, A. Gandhi, S. Patil, and O. Deshmukh, "Livelinet: A multimodal deep recurrent neural network to predict liveliness in educational videos." *International Educational Data Mining Society*, 2016.

[136] W. Liu, Z. Wang, X. Liu, N. Zeng, Y. Liu, and F. E. Alsaadi, "A survey of deep neural network architectures and their applications," *Neurocomputing*, vol. 234, pp. 11–26, 2017.

[137] S. Samarasinghe, *Neural networks for applied sciences and engineering: from fundamentals to complex pattern recognition.* Crc Press, 2016.

[138] B. Mehlig, "Artificial neural networks," *arXiv preprint arXiv:1901.05639*, 2019.

[139] R. Yamashita, M. Nishio, R. K. G. Do, and K. Togashi, "Convolutional neural networks: an overview and application in radiology," *Insights into imaging*, vol. 9, no. 4, pp. 611–629, 2018.

[140] H. Salehinejad, S. Sankar, J. Barfett, E. Colak, and S. Valaee, "Recent advances in recurrent neural networks," *arXiv preprint arXiv:1801.01078*, 2017.

[141] S. Zheng, K. Ristovski, A. Farahat, and C. Gupta, "Long short-term memory network for remaining useful life estimation," in *2017 IEEE international conference on prognostics and health management (ICPHM).* IEEE, 2017, pp. 88–95.

[142] J. Wang, K. Wang, Y. Wang, Z. Huang, and R. Xue, "Deep boltzmann machine based condition prediction for smart manufacturing," *Journal of Ambient Intelligence and Humanized Computing*, vol. 10, no. 3, pp. 851–861, 2019.

[143] F. Farahnakian and J. Heikkonen, "A deep auto-encoder based approach for intrusion detection system," in *2018 20th International Conference on Advanced Communication Technology (ICACT)*. IEEE, 2018, pp. 178–183.

[144] G. Liu and J. Guo, "Bidirectional lstm with attention mechanism and convolutional layer for text classification," *Neurocomputing*, vol. 337, pp. 325–338, 2019.

[145] W. Yu, X. Sun, K. Yang, Y. Rui, and H. Yao, "Hierarchical semantic image matching using cnn feature pyramid," *Computer Vision and Image Understanding*, vol. 169, pp. 40–51, 2018.

[146] T. Fischer and C. Krauss, "Deep learning with long short-term memory networks for financial market predictions," *European Journal of Operational Research*, vol. 270, no. 2, pp. 654–669, 2018.

[147] M. A. Mohammed, M. K. Abd Ghani, N. a. Arunkumar, R. I. Hamed, S. A. Mostafa, M. K. Abdullah, and M. Burhanuddin, "Decision support system for nasopharyngeal carcinoma discrimination from endoscopic images using artificial neural network," *The Journal of Supercomputing*, vol. 76, no. 2, pp. 1086–1104, 2020.

[148] L. Deng, D. Yu, and G. E. Dahl, "Deep belief network for large vocabulary continuous speech recognition," Mar. 3 2015, uS Patent 8,972,253.

[149] K. Sabanci, A. Kayabasi, and A. Toktas, "Computer vision-based method for classification of wheat grains using artificial neural network," *Journal of the Science of Food and Agriculture*, vol. 97, no. 8, pp. 2588–2593, 2017.

[150] T. Le, P. Pardo, and W. Claster, "Application of artificial neural network in social media data analysis: a case of lodging business in philadelphia," in *Artificial Neural Network Modelling*. Springer, 2016, pp. 369–376.

[151] P. P. Bendangnuksung, "Students' performance prediction using deep neural network," *International Journal of Applied Engineering Research*, vol. 13, no. 2, pp. 1171–1176, 2018.

[152] G. Zhong, L.-N. Wang, X. Ling, and J. Dong, "An overview on data representation learning: From traditional feature learning to recent deep learning," *The Journal of Finance and Data Science*, vol. 2, no. 4, pp. 265–278, 2016.

[153] S. Umair and M. M. Sharif, "Predicting students grades using artificial neural networks and support vector machine," in *Encyclopedia of Information Science and Technology, Fourth Edition*. IGI Global, 2018, pp. 5169–5182.

[154] D. Charte, F. Charte, M. J. del Jesus, and F. Herrera, "A showcase of the use of autoencoders in feature learning applications," in *International Work-Conference on the Interplay Between Natural and Artificial Computation*. Springer, 2019, pp. 412–421.

[155] L. P. Perera and B. Mo, "Ship performance and navigation data compression and communication under autoencoder system architecture," *Journal of Ocean Engineering and Science*, vol. 3, no. 2, pp. 133–143, 2018.

[156] T. Dumas, A. Roumy, and C. Guillemot, "Autoencoder based image compression: can the learning be quantization independent?" in *2018 IEEE International Conference on Acoustics, Speech and Signal Processing (ICASSP)*. IEEE, 2018, pp. 1188–1192.

[157] S. Xie, J. Gu, D. Guo, C. R. Qi, L. Guibas, and O. Litany, "Pointcontrast: Unsupervised pre-training for 3d point cloud understanding," in *European Conference on Computer Vision*. Springer, 2020, pp. 574–591.

[158] K. Simonyan and A. Zisserman, "Very deep convolutional networks for large-scale image recognition," *arXiv preprint arXiv:1409.1556*, 2014.

[159] A. Krizhevsky, I. Sutskever, and G. E. Hinton, "Imagenet classification with deep convolutional neural networks," *Communications of the ACM*, vol. 60, no. 6, pp. 84–90, 2017.

[160] A. Tato, R. Nkambou, A. Dufresne, and M. H. Beauchamp, "Convolutional neural network for automatic detection of sociomoral reasoning level." *International Educational Data Mining Society*, 2017.

[161] M. Mezzini, G. Bonavolontà, and F. Agrusti, "Predicting university dropout by using convolutional neural networks," in *INTED2019 Proceedings*. IATED, 2019, pp. 9155–9163.

[162] F. Okubo, T. Yamashita, A. Shimada, and H. Ogata, "A neural network approach for students' performance prediction," in *Proceedings of the seventh international learning analytics & knowledge conference*, 2017, pp. 598–599.

[163] F. Okubo, T. Yamashita, A. Shimada, and S. Konomi, "Students' performance prediction using data of multiple courses by recurrent neural network," in *25th International Conference on Computers in Education, ICCE 2017*. Asia-Pacific Society for Computers in Education, 2017, pp. 439–444.

[164] C. Lin and M. Chi, "A comparisons of bkt, rnn and lstm for learning gain prediction," in *International Conference on Artificial Intelligence in Education*. Springer, 2017, pp. 536–539.

[165] E. Lau, L. Sun, and Q. Yang, "Modelling, prediction and classification of student academic performance using artificial neural networks," *SN Applied Sciences*, vol. 1, no. 9, pp. 1–10, 2019.

[166] M. Sarkar and A. De Bruyn, "Lstm response models for direct marketing analytics: Replacing feature engineering with deep learning," *Journal of Interactive Marketing*, vol. 53, pp. 80–95, 2021.

[167] G. Chen, D. Ye, Z. Xing, J. Chen, and E. Cambria, "Ensemble application of convolutional and recurrent neural networks for multi-label text categorization," in *2017 International Joint Conference on Neural Networks (IJCNN)*. IEEE, 2017, pp. 2377–2383.

[168] A. Sherstinsky, "Fundamentals of recurrent neural network (rnn) and long short-term memory (lstm) network," *Physica D: Nonlinear Phenomena*, vol. 404, p. 132306, 2020.

[169] C. N. Duong, K. Luu, K. G. Quach, and T. D. Bui, "Deep appearance models: A deep boltzmann machine approach for face modeling," *International Journal of Computer Vision*, vol. 127, no. 5, pp. 437–455, 2019.

[170] B. J. Fogg, "A behavior model for persuasive design," in *Proceedings of the 4th international Conference on Persuasive Technology*, 2009, pp. 1–7.

CPSIA information can be obtained
at www.ICGtesting.com
Printed in the USA
LVHW082006250622
722129LV00009B/144

9 787877 800049